MUIRHEAD'S MYSTERIES

2009 - 2010
Leaves from a Fortean Zoologist's Notebook

Volume One

Typeset by Jonathan Downes, Chloe Gray
Cover and Layout by SPiderKaT for CFZ Communications
Using Microsoft Word 2000, Microsoft Publisher 2000, Adobe Photoshop CS.

First published in Great Britain by CFZ Press

CFZ Press
Myrtle Cottage
Woolsery
Bideford
North Devon
EX39 5QR

ISBN: 978-1-909488-48-9

Dedicated to my Mother, Sheila Muirhead, and Family

Dragons Are Too Seldom

*To actually see an actual marine monster
Is one of the things that do before I die I wonster.
Should you ask me if I desire to meet the bashful inhabitant of Loch Ness,
I could only say yes.
Often my eye with moisture dims
When I think that it has never been my good fortune to gaze on one of Nature`s whims.
Far from ever having seen a Gorgon
I haven`t ever seen the midget that sat in the lap of Mr Morgan…*

Ogden Nash (1935)

ACKNOWLEDGEMENTS

I would like to thank my parents for encouraging and nurturing my interest in wildlife from a very early age. My father, Stuart Muirhead, mentioned to me once that he found me playing with a couple of worms in the car park of our block of flats in Hong Kong, when we were living there in the 1970s. These were the common earth worm variety, not the Mongolian Death Worm displaced to another part of Asia! My father also used to send me first day covers of Hong Kong's colourful bird stamps whilst I was at boarding school in England. I'd like to thank the rest of my family for accepting this "zoological anarchist" into their midst, with barely a glazed eye amongst them. I'd like to thank Devo for my impetus to question zoological orthodoxy just as they question social and political orthodoxy. I would also like to thank Jon Downes, Chloe Gray and the rest of the CFZ crew, and Dr Karl Shuker for his very generous introduction. I would like to thank others who have influenced me: Charles Fort, Bernard Heuvelmans, William Corliss, Bob Skinner, the late Jan Williams (without whom I may never have got back in contact with Jon after so many years), Corinna Downes for her mastery of bullets, Kay Coggin, Loes Modderman, Rebecca Lang, Lizzy Bitakaramire, Paul Sieveking, Dr Darren Naish, Richard Freeman, Marcus Matthews, Paul Screeton, Chad Arment, Francois de Sarre and any other cryptozoologists and Forteans whose paths (ley lines?) I've crossed. I would like to thank my mother Sheila Muirhead for doing some proof-reading. I would like to thank the ever-patient John Smith of the Disability Information Bureau in Macclesfield, who hails from the land of Nessie, for helping me work on the text of Muirhead`s Mysteries two or three years ago. I would like to thank the management of the Snowgoose bar in Macclesfield for allowing me to use their premises to have the photos of Dr Devo on the front and back cover of this book. Last but not least, the large anonymous group of journalists, librarians, archivists and eye witnesses who have collected data recorded in this book.

FOREWORD
BY Dr KARL SHUKER.

The fourth book in Charles Fort's classic, seminal quartet dealing with anomalous phenomena is *Wild Talents*, first published in 1932, which dealt with a wide range of extraordinary mental powers or talents apparently exhibited by certain humans that went far beyond the norm. If there is such a phenomenon as a 'wild talent' for bibliographical research, then Richard Muirhead definitely possesses it, because in my opinion he has no equal among the Fortean community in terms of his peerless ability to uncover the most esoteric yet invariably captivating, hitherto-cryptic nuggets of information relating to mysteries and curiosities of every imaginable – and quite often wholly unimaginable – kind.

Nor does he even risk life and limb journeying through the wilder realms of the world in search of such treasure – most of his quests for arcane material are conducted in the comfortingly tame surroundings of his local public library in Macclesfield, Cheshire, in England. Yet his finds are invariably just as exotic and entrancing as anything disinterred in far-flung lands of the lost by teams of ardent archaeologists or other dedicated delvers into the silent, shadowy vaults of the past.

Consequently, it was only a matter of time before Richard committed a collection of his most memorable bibliographical discoveries to print, and now I am delighted to say that he has done so. Harvested from his regular, long-running online column 'Muirhead's Mysteries' contained within the *CFZ Bloggo*, in this fascinating volume – which I sincerely hope will be just the first of many in a continuing 'MM' series – Richard has surveyed and selected a truly eclectic array of cryptozoological and other animal-related enigmas with which to bemuse and bedazzle his readers.

Unlike so many of his peers, however, he has not wasted time labouring over the tried and trusted subjects invariably included in treatments dealing with unexplained creatures. Instead, I can guarantee that virtually everything waiting to astonish you within the pages of this unique book will either be entirely novel, untarnished by previous retellings in earlier books, or will be familiar to you only because it was in Richard's online column that these stories first attracted recent attention. True, many of the reports date back very considerably in time, but because they have somehow eluded other Fortean researchers, having slipped through the cracks in the paving stones of widespread public attention until spotted by this present book's eagle-eyed author and rescued from longstanding obscurity, they remain as fresh and as thought-provoking today as they were when first penned by their long-forgotten original scribes.

Indeed, where else can you possibly read about the anomalous marine seahorses allegedly existing in Bolivia's freshwater Lake Titicaca, gargantuan eagles and thunderbird nests, hairy tortoises and feathered fishes, snake stones in Yorkshire and snakes with legs (one of Richard's specialities), giant ground sloths in New Zealand, paradoxical pygmy weasels, jackals in Britain and some very Fortean foxes, blood-drinking butterflies and caterpillars of the Kaiser, Ireland's extraordinary battle of the starlings and Limerick Cathedral's monsters of the misericords, incongruous insects and other interesting invertebrates, super rats and suicidal mice, the (literally!) stomach-churning bosom serpent, giant lizards in New Zealand, the dragons of Sussex and Sweden, whistling trees, weeping trees, and even walking trees, a leopard of unknown origin shot in the wilds of Britain in 1836!, sheep-sized mystery birds in Uganda, monstrous dragonflies, singing spiders, and entombed scarlet bats, a hitherto-overlooked whale-clubbing sea monster from Alaska, the ikanda or bear-ape of Equatorial West Africa, a hot-water octopus, the enigmatic dolt-cat, a putative living dinosaur in Idaho, South America's dreadful devil-sticker, crocodiles and a dingo in Ireland, a transparent man-ape and tiger-striped wolves, some decidedly odd otters, a shrew that was blue, flying snakes of Manchuria, death by cryptid, a veritable menagerie of mysterious beasts from Hong Kong (where Richard lived as a child) and a Macclesfield jackalope, even some delightful self-penned poems about unicorns, tadpoles, mermaids, and other intriguing entities, plus so much more besides that my poor data-bombarded mind becomes feverish even attempting to encompass them, let alone recall them all??

So put away your laptop and your tablet, disconnect yourself from your ipod and your smartphone, and tarry a while instead amid the sequestered, secluded world of Muirhead's Mysteries, delighting in the inexplicable fauna (and also some flora) of fancy, fantasy, folklore, and fact that can only be encountered within this wonderful book. You know it makes sense!

Dr Karl P.N. Shuker, October 2016.

INTRODUCTION

I was born in Hong Kong in 1966 and therefore was surrounded by some of the creatures mentioned within this book and probably others who make it into these pages as cryptids. When I was ten, around about April 25th 1977, the news that fishermen on the Japanese trawler Zuiyo Maru had dragged up from the depths of the South Pacific, a carcass of a dead marine animal that many cryptozoologists and mainstream marine biologists insist was a basking shark (are they trying to tell us the marine biologist onboard, the very experienced Michihiko Yano, was completely wrong and couldn't even identify a species of shark with which he was probably familiar?) but I believe was a plesiosaur, was reported in the Hong Kong *South China Morning Post.* From then on the stream of my mind ran into cryptozoology and later, in the early 1980s I discovered Charles Fort and the *Fortean Times.*

Between about 1972 and 1994 I lost touch with Jon Downes, who also lived in Hong Kong, then in the first of many serendipitous events , whilst I was a student in Cork, I somehow was able to contact Jan Williams, then editor of a long defunct publication called *SCAN*. She put me in touch with Jon who was about to launch *Animals and Men*. The rest, as they say, is crypto-history.

"Come in cryptozoologist, you're late! Have you got the book China man gave you?!"..."Yes readers, in the past this information in Muirhead's` Mysteries has been suppressed, but now it can be told: Every man, woman and mutant on this planet shall know the truth about the inner core of anomalous zoology!" (Adapted from Devo`s video for Jocko Homo.) [1]

Deciding to begin a blog on cryptozoology about six years ago reminds me that the world is much, much stranger than most non-Forteans would believe. Which is hardly surprising to say given the massive resource that is the World Wide Web, a tool that has opened up a strange world that's always been full of anomalies. It's not necessarily that the world is getting stranger or that people are getting odder, but the world from a zoological point of view has always thrown up oddities. (In a world where even computers are evolving of their own, according to a story in the late William Corliss's Science Frontiers no. 108, November-December 1996,where Corliss notes an occasion about 60 years ago when a first generation vacuum-tube-equipped MANIAC I computer was resourceful and innovative enough to "create" an un-programmed chess move against S. Ulam, whom it was playing against, then anything must be possible in the natural world.) I am prompted to consider the fact that I wish I was a much more committed zoologist. I decided to Google Image the words "zoologist" and "cryptozoologist" (in early February 2015) and the gap between the subject matter was too profound. I believe these disciplines in fact overlap. Which came first, the chicken or the egg? In this case the chicken is zoology and the egg is cryptozoology.

As an afterthought I guess I'd describe myself as a "zoological anarchist". I must admit I am disappointed on looking out of my window not to see beavers or wildebeest frolicking in the middle distance which brings me neatly onto my concentric circles theory. This is; that in the outer core of "reality" (whatever that is), we have zoology, then within the next circle, cryptozoology, then finally, within the innermost circle, the "high strangeness" of cryptozoology, where Muirhead's Mysteries, lies, or hibernates, and where you might find, if not a beaver then something almost as unexpected. Of course these circles overlap.

There's an occasion during Devo's performance in August 1996 whilst they're playing in support of Metallica at a gig in Irvine Meadows, California, when Gerald V. Casale, Devo's vocalist, synthesizer play and bass guitarist cries out during Jocko Homo, "I think de-evolution is real, just look around!" Well as the author of Muirhead's Mysteries, I want to echo Casale's and state "I think Muirhead's Mysteries are real, just look around"; the world, or that small corner of the world you inhabit and not just around but within this book of my blogs covering the disciplines of cryptozoology, Fortean zoology, folklore and the paranormal, from my home county of Cheshire, U.K. to my birth place of Hong Kong, China and beyond.

The following comment could easily apply to our activities as cryptozoologists and the endeavour to uncover the planet's remaining mysteries:

> "As the Chinese proverb goes, the nail that sticks out gets pounded. You are especially scrutinized and judged if your success carries any whiff of change or controversy. If you challenge authority in any real manner, your voice in the marketplace is quickly revoked." : Gerald V. Casale `Watch Us Work It`: A reflection on "work" by Devo founder, Gerald V. Casale (2011) Essay, published 2011-06-26, in *Essay Collection* at Workstew.com. [2]

Back in the very early days of my blogs, specifically on March 26th 2009, Jon gave a story of a Northern Irish sea serpent the memorable title `How The Bloody Hell Does Muirhead find this stuff? `. I joked "it's the hand of God and the fingers of Richard Muirhead" mis-quoting the (in?) famous statement by Argentinian footballer Maradona in 1986 concerning that goal against England. The truth is somewhat more transparent, but not by much. The stories seem to find me, in a serendipitous sense. I may be de-evolving myself, back to being a young frolicking primate, (if life gets too difficult I can always go back to the jungles of The Peak, Hong Kong island or Macclesfield Forest and swing in the trees) but the things I find and put in the blog are doing the opposite! I mean evolving. By random mutations? Who can tell?!

I would like to thank the Dear Leader himself (Jon Downes has had to put up with a lot of silliness from me, including my totally frivolous and surreal comparison between himself and the dictator of North Korea) for publishing this collection and all you out there who have read, and may now begin to read, Muirhead's Mysteries. As Bernard Heuvelmans has said "there are lost worlds to be discovered in your back garden." Jon, I think would say, "Onwards and Upwards!" As Devo would sing in `An Uncontrollable Urge` and I would humbly suggest regarding this book:

"It's got style, its got class
So strong, I can't let it pass
I gotta tell you all about it
I gotta scream and shout it." [3]

I have an uncontrollable urge to tell you about a giant dragonfly in Yorkshire, an entombed lizard in a brick in Macclesfield and a giant sturgeon in Hong Kong, but that will have to wait for a future volume of Muirhead's Mysteries! I am about to tell you about giant bird's nests in the Nevada desert, a fictitious amber coloured Irish tadpole and strange sea horses in Lake Titicaca, so please read on!

And finally I also wish to echo Fort: "I am a collector of notes upon subjects that have diversity — such as deviations from concentricity in the lunar crater Copernicus, and a sudden appearance of purple Englishmen — stationary meteor-radiants, and a reported growth of hair on the bald head of a mummy — and 'Did the girl swallow the octopus?'" [4]

Given that I am the man that suddenly went blond to look like a very much alive Andy Warhol, I hope I'm qualified to present to you, dear reader, this book.

Thanks Jon and Enjoy!
September 10th 2016

REFERENCES

1. Adapted from a video of Devo *Jocko Homo*: https://www.youtube.com/watch?v=hRguZr0xCOc&safe=active
2. http://en.wikiquote.org/wiki/Devo
3. Devo: Uncontrollable Urge lyrics.
4. Charles Fort `Wild Talents`. The Complete Books of Charles Fort. Dover edition (1974) page 846.

MONDAY, JANUARY 26, 2009

MUIRHEAD'S MYSTERIES: The mysterious seahorses of Lake Titicaca

Preamble by Jon: When I was a child I lived in Hong Kong. I spent the latter part of August and early September 1970 in hospital undergoing a barbaric operation on my knees. I went in on the day after my eleventh birthday, and stayed in for the next four weeks.

However, I was allowed to come home for weekends, and every Friday afternoon an ambulance would drive from Queen Mary Hospital down in Pokfulam, up the Peak, until I was carried into our ground floor flat at Peak Mansions by two burly young ambulancemen. Somehow, I can surmise with the benefit of hindsight, that my father had pulled rank, because I had even been provided with a hospital bed on wheels, (something which I doubt was a service available to all and sundry) and on very sunny days Ah Tim and Ah Tam (the Chinese couple who looked after us) would wheel me out into the conservatory, and open the French windows so I could haul myself up in bed using the support bars, and look out onto the world outside.

Peak Mansions was (I use the past tense because it was apparently demolished in 1989) a six story squat building with an impressively mock Georgian façade, and two seemingly pointless green domes on the roof. It had originally been built either late in the 19th or early in the 20th century as accommodation for expat Civil Servants, but during WW2 it was the home of the Hong Kong Volunteer Force and was badly damaged by shelling. However, it was seriously rebuilt after the war, and carried on with its original purpose. Running along the front of the building was Peak Road, and on the other side of the road was a heavily forested hillside which tumbled down for miles to the town (now city) of Pokfulam. This forest was the haunt of leopard cats, pangolins, civets, and porcupines, and within living history had been home to tigers and possibly even leopards. It was a magnificent place, and I spent much of my childhood exploring it, and much of my adulthood dreaming about it, and after a week of surgery (they botched the operation the first time and had to do it again) and physiotherapy, just to lie in my bed looking through the open French windows at the jungle below was bliss.

Outside the windows, a shallow sloping stretch of lawn led down to the road, and my mother was wont to recline on a sun lounger there and sunbathe. Occasionally she would be joined by her friends, and on this particular occasion a lady called Sheila Muirhead, with an irritating young son aged four had come to visit. I was annoyed. Now my mother would not be willing to tell me stories, or make too much of a fuss of me, and what was worse, her son was too young for me to be able to talk to on any meaningful level, and as both my legs were in plaster, and I was wracked with agony every time I moved, I couldn't do anything more boisterous in terms of play.

Then I had an idea.

For my birthday, the day before I went into hospital I had received a copy of *Hong Kong Butterflies* by Major J.C.S Marsh, and I was desperate to put my newfound book to use. I was at the age when I had just begun to realise that some creatures were more closely related than others, and I wanted to identify the myriad animals that surrounded me. Fluttering along a few inches above the closely mown grass were dozens of small, blue butterflies. Major Marsh listed several dozen members of this family, quite a few of which looked very similar.

So I called to the toddler who was earnestly chuffing up and down the sloping lawn pretending to be a goods train.

"Hello," I said. "I'm Jonathan. What's your name?"

"Richard," he said. "What are you doing in bed?"

So I told him, and despite the seven year gap in our ages, he not only seemed to sympathise with me, but – after I explained my predicament re. Major Marsh and the blue butterflies - he expressed - as well as a four-year-old can express anything – a willingness to help me in my investigations. So I told him where my bedroom was, and where I kept my butterfly net, and where my precious copy of *Hong Kong Butterflies* was, and he trotted off inside. About ten minutes later, after a few false starts, I was sitting up in bed with Major Marsh's magnum opus on my knee, and a Robinson's marmalade jar in one hand, as my young assistant – still making enthusiastic train noises – rushed up and down in search of butterflies.

We have been friends ever since. Richard has what Charlie Fort would have no doubt described as a "Wild Talent". He is the best researcher I have ever met, and invariably has something interesting on the go. I asked him if he wanted to write for Cryptozoology Online (I really must stop calling it the CFZ bloggothing) and he immediately asked me if I knew anything about mysterious freshwater seahorses in South America…

Seahorses are a genus (Hippocampus) of fish belonging to the family *Synguathidae*, which also include pipefish and leafy sea dragons.

Fig. 2 (tamaño natural)

H. pxcampus titicacensis

There are over 32 species of seahorse, mainly found in shallow tropical and temperate waters throughout the world. They prefer to live in sheltered areas such as sea grass beds, coral reefs, or mangroves….The male sea horse can give birth to as few as 1 and as many as 2000 "fry" at a time and pregnancies last anywhere from two to four weeks, depending on the species." [1]

They must be amongst the most interesting of sea creatures. There are many different species all over the world. Whilst reading William Corliss's seminal work *Anomalies in Geology*, on pages 53-54,in the chapter ESB6 `Living Organisms and Recent Fossils Found At High Altitudes` I was very surprised to discover reports of living seahorses in Lake Titicaca.

Lake Titicaca, which straddles the Bolivia-Peru border at a height of 12,530feet/3820 m above sea

level, is an unusual place. Not only do the local inhabitants use reed boats to traverse the lake, similar to the ones used in ancient Egypt ["In fact, the type of reed plant is reportedly the same as the reed plant used in ancient Egypt..." [2]] but up to at least the end of World War 2 and may be later, a species of sea horse, *Hippocampus titicacensis* was said to inhabit the Lake, proving to some that the lake was once attached to the Pacific Ocean. Corliss refers to beliefs that the Andes may be very "young" geologically and that the putative Lake Titicaca seahorse, very high beaches with recent marine shells and recent plant fossils at very high altitudes are a sign of this.

There is a photo of a degenerate (because of in-breeding?) seahorse from the lake on the website of Jim Forshey and Bruce Watts, two American experts on seahorses.

This photo dates from about 1943 and is said to be a specimen at the Museum of Natural History in Berlin. [3] Arthur Posnansky, who mentions this seahorse in his volume one of *"Tihuanacni: Cradle of American Man"* stated: "The Indians believe that this animal is a sort of divinity, like everything that is strange and inexplicable to them."[4] what is really inexplicable is, how did the seahorse get into the Lake and where are they now?

Even in the 1940s they were very rare and very recent web sites do not mention them at all, though trout, "suche" and "capache," flamingos and duck are mentioned. If anyone can shed light on these enigmatic seahorses please can they contact the CFZ?

Thank you.

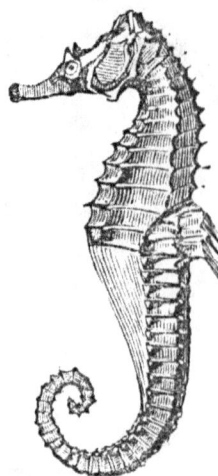

REFERENCES.

1. Project Seahorse. The Biology of Seahorses. Reproduction. In http://en.wikipedia.org/wiki/Seahorse
2. Donald E. Chittick *The Puzzle of Ancient Man* (Newberg, Oregon), p193
3. See: http://www.seahorses.com/ AquariumAndFishItemsForSale/ BibliographySeahorses2005Edition.pdf
4. Ibid.

WEDNESDAY, FEBRUARY 4, 2009

The Ropen - a living pterosaur in Melanesia?

If there is any possibility of prehistoric survivors on the planet then Melanesia and Papua New Guinea would be the place where they might live. The Ropen ("demon flyer") a putative pterosaur, "said to frequent lakes and mountain caverns" according to M. Newton`s *Encyclopaedia of Cryptozoology* [1] was brought to the attention of the cryptozoological world in late 1999 by Bill Gibbons.[2] It is actually quite surprising that the Ropen is still not particularly well known in the cryptozoological world given that it is ten years since its "discovery", though this is changing with the publication of the 2nd edition of Jonathan Whitcomb`s *Searching for Ropens* in 2007.

See also http://www.ropens.com/ There are also its parallels elsewhere in the world which should be considered. For example, the giant flying snake of Namibia is said by locals to have a light on its head [3.] Could this flying snake actually be a pterosaur? It is possible. Roy Mackal in his book *Searching for Hidden Animals* mentions reports of living members of "extinct pterodactyls" in Kenya in March or April 1974 and in a swampy area of Namibia in late 1975, allegedly filmed by an American team, but this is nearly all the information Mackal gives. [4.] Large flying creatures are also reported from Mexico and parts of the U.S. In Indonesia, the ahool and the Orang-bati may relate to the Ropen of Umboi Island off Papua New Guinea.

One of the earliest reports of the Ropen was from Perth, Australia, rather than Papua New Guinea. In December 1997, one evening, a couple saw a flying creature with "a lizard appearance" and a size of "thirty to fifty feet". Even earlier, in 1944, Duane Hodgkinson, a friend and a native guide witnessed a giant flying pterodactyl emerge from the undergrowth having been startled by a pig. "The creatures` [sic] colour was dark, its neck was long, and the head-length…Both the head-crest and beak (or mouth) were long and narrow, parallel to each other, and, in flight, almost parallel to the neck" [4] In 2003 a man named Abram saw a Ropen glowing red and white flying 100m above Opai Beach on Umboi Island. The bioluminescence is said to be related to secretions which fall from the creatures as they fly, which can cause severe burns to human skin. In October (2004?) a David Woetzel saw a golden shimmering light coming from the direction of Mount Barik, on Umboi Island, flying at mountain top height. It has been said that the Ropen light is brighter than the moon. Frighteningly, the Ropen was once said to rob human bodies from graves. It is worshipped by at least some in parts of Papua New Guinea and there are said to be grey-spotted, blue-spotted and dark-spotted Ropens. Critics of the Ropen as pterosaur theory suggest that the Ropen is a flying fox bat. But as Whitcomb points out: "……fruit bats never grow long tails, never eat fish, never glow at night, and never dig up the grave of a recently-deceased human to carry away the body." [5]

REFERENCES.
1. M. Newton. Encyclopaedia of Cryptozoology. p.401
2. Ibid p.401
3. R. Muirhead. The Flying Snake of Namibia: An Investigation. In: Centre for Fortean Zoology Yearbook 1996. pp. 112-123
4. R. Mackal. Searching for Hidden Animals. p.54
5. J. Whitcomb Searching for Ropens. 2nd edition p.26

TUESDAY, FEBRUARY 10, 2009

Craziness and Cryptozoology in Northampton, 1997

As regular readers know, Richard is one of my dearest friends... What people may not know is that we have something else in common - we both suffer from mental health issues - and in Richard's case his problems were so bad that they landed him in hospital. I have always been open about my own battle with a bipolar illness, and I would just like to say how proud I am of Richard for `coming out` on such a public forum as this. Well done mate...

One fine mid-summer afternoon in early August 1997 I was sitting at a computer in the Bodleian Library in Oxford, doing research for my MA dissertation. As I was sitting there a strange series of thoughts entered my head and I ended up believing I was the Antichrist. (I am not, I am a Christian.) So this delusion got worse and worse until that evening I was put in an ambulance, sedated, and taken to the main psychiatric hospital in Northampton (because there were no vacancies in Oxford). Coincidentally, this is where the "peasant poet" John Clare, (below left) more famous for his nature poetry, was also a patient.

His harrowing poem `I am` about the alienation from those around him that his illness brought to him has inspired me down the years.

I am: yet what I am none cares or knows,
My friends forsake me like a memory lost;
I am the self-consumer of my woes,
They rise and vanish in oblivious host,
Like shades in love and death's oblivion lost;
And yet I am, and live - like vapours tossed

However, I digress.

So I arrived and almost immediately there were cryptozoological stories to investigate. There was a picture above my hospital bed of a vase of flowers, a reprint of the 18th century horticulturalist Robert Furber's *Twelve Months of Flowers*, 1730, picture for March. At the bottom, beneath the vase was a pair of animals that looked like a kind of Tatzelwurm. Jon discussed it with the other members of the CFZ, then at Exeter, and they decided it was an early depiction of a dolphin.

Whilst in hospital (which was luxurious by the way) I met a woman and we began to talk about black squirrels in Britain.

She gave me the address of a female friend she had, who lived in Cambridgeshire. Moreover this lady had photos of black squirrels. Sure enough, when I wrote to her she passed on one of her photos which were later published in *British Wildlife*, December 1998 along with my essay on the subject.

After a week or two in the psychiatric hospital, I was allowed to go into Northampton accompanied by a nurse. I was in Northampton until early September, until shortly after the death of Princess Diana in Paris.

On one occasion I went to the local studies library and in one or two days it turned out to be the most productive, or - at least - *one* of the most productive short term periods of cryptozoological research in my life. The library was a delightful old building, and the animal stories below were found by using a card catalogue system. As a qualified librarian I was taught that with the advent of information technology, this is not considered the most time efficient method of searching! But I can see no harm.

Below, in date order, are most of my unusual animal stories from Northampton. A few have been left out or abbreviated, due to their comparatively uninteresting nature.

- Horned horse. *Northampton Mercury.* June 27th 1795. Nearly all the cuttings are abbreviated "N. Merc", so I presume that means "Northampton Mercury."

When I first read those words "Horned Horse" I first thought "is this some kind of unicorn?" Then I found out this was actually a gnu. "The following singular accident took place about a fortnight since, on the Chester road, near the sand house, a short distance from Brick-hill. A caravan in which were a Leopard, a Bison (or Buffalo) a Horned Horse, and several smaller wild beasts, owing to the axle-tree breaking, was pitched on one side....etc.*"*

- Locusts *Northampton Mercury.* September 26th 1857.

This is one of my favourite entomological anomalies. A locust was found in Adam and Eve

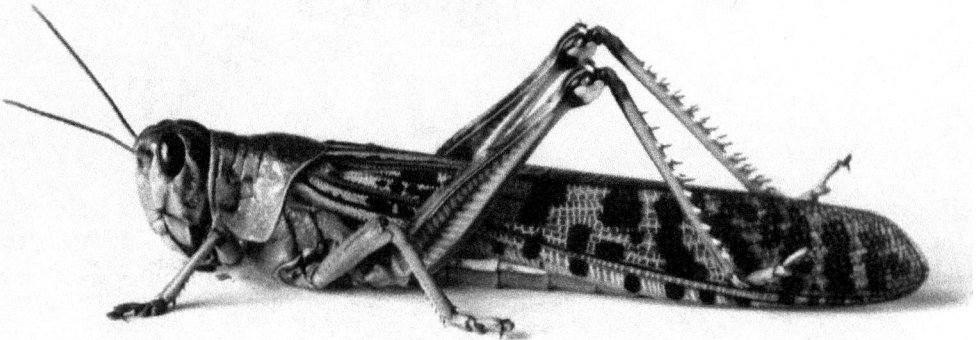

Street, Market Harborough on Sunday last, while the Rev. J Clifford was preaching in the open air. A servant in the employ of Mr H. Foster took it home with him; it measures more than two inches in length. One was seen on Friday last in a field of turnips in the parish of Cobourn, near Caistor. Its presence was known by a skip of four or five yards, and on arresting the attention of the beholder impressed him at first that it was a small rising from the ground. A locust was found in St Paul's Street, Stamford, a few days ago; a dog was playing with it at the time. It survived several days." Note the coincidences: Adam and Eve and St. Paul. Locusts in the Bible are a symbol of God's judgement.

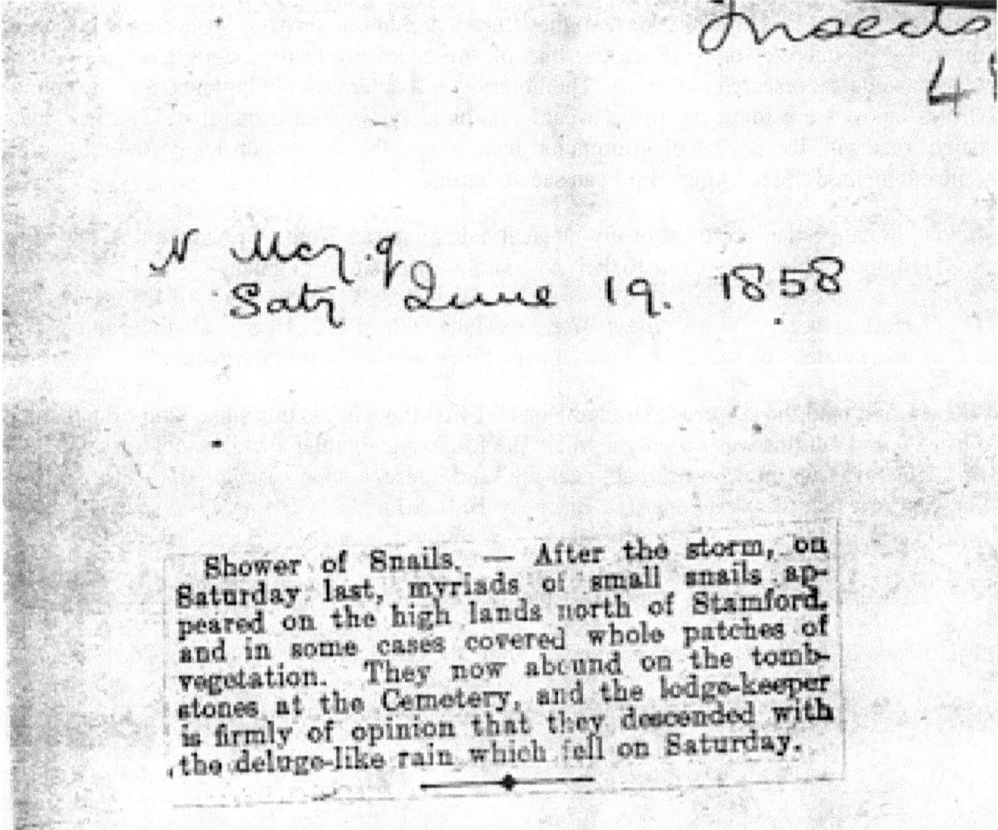

- Shower of snails. *Northampton Mercury*. June 19th 1858.

"Shower of Snails-After the storm, on Saturday last, myriads of small snails appeared on the high lands north of Stamford and in some cases covered whole patches of vegetation. They now abound on the tombstones at the cemetery, and the lodge-keeper is firmly of the opinion that they descended with the deluge-like rain which fell on Saturday."

Fort mentions a fall of snails in *The Book of The Damned*: In Cornwall, near Redruth, on July

8th 1886.This is in *Science Gossip* 1886-238. [1]

- Locust *Northampton Daily Chronicle* (?) August 22nd 1901

"Another locust, this time a South African one: A fine specimen of the South Africa locust was found in Derngate, Northampton, by Mr Frank Bex this morning. Mr Bazeley, Sheep Street has it on view in his window."

- An Indian mongoose. *Northampton Daily Record* (?) September 28th 1904

"A Rare Animal. Whilst out rabbit-shooting on Tuesday last week, at Mr Banks` lime kilns, near Moulton Park. Mr W. Parbery, of the Old Five Bells Kingsthorpe, was fortunate enough to secure a rare natural history specimen. There was a good deal of speculation as to what the animal was but through the kindness of Mr T. J. George and others it has been found out to be an Indian mongoose. The animal, when "bolted" by a ferret (which was muzzled) showed fight, but was soon despatched. It is supposed that the specimen (which is being preserved) had escaped from confinement."

- Cat feeding chick. *Northampton Daily Chronicle*. September 23rd 1905.

"Anomaly in Natural History." Your readers may be interested in the following anomaly in natural history: Mrs Jackson (wife of Mr Smyth`s gamekeeper) put a lame chicken into the basket of her cat with one kitten to nurse. The cat took kindly to her charge, keeps it warm under her; treats it exactly as she does her kitten, licking it clean, etc., and if the chick is taken out of the basket carries it back in her mouth. The chick is doing well, and will soon be independent of its strange fostermother ". J. T. BARTLETT.

- A gazelle. *Northampton Mercury*. May 14th 1909.

This gazelle was supposed to have escaped from a travelling menagerie. This is the usual explanation given in early 20th century British newspapers for exotic animals found or shot in the countryside. But in this story about a gazelle there is a twist: The gazelle was thought to have been being suckled by a cow belonging to a farm owner, Mr Charles Kingston! I will quote from parts of the cutting: My own comments in brackets.

"No little excitement was caused in the village (of Caldecote) on Monday (i.e. the 10th) when it became known that a young gazelle had been shot on Uplands Farm. It appears that for some considerable time the occupier of the farm, Mr Charles Kingston, had been at a loss to understand the reason why one of his best milk cows was yielding so little milk....Mr Kingston, on receiving information, set off with his man and two lads to stalk the intruder, and after an exciting time got in a favourable position for a shot and brought his quarry down. Several people who have been in South Africa are of opinion that the animal is a young gazelle. How it came to Caldecote can only be conjectured, but the field where it was found lies close to the main road, and it is possible that it may have escaped from a passing menagerie......Judging by the time when the milk cow unaccountably failed to yield her usual supply of milk Mr Kingston thinks that the gazelle must have been on the farm for the greater part of the winter and suckled by the cow."

- Another locust. *Northampton Mercury*. September 24th 1909

An Unwelcome Visitor - A remarkably fine specimen of the locust was captured on Thursday, in a bean field belonging to H. B. Whitworth, Esq., near the Northampton toll-gate, on the Kettering-road, by a lad named John Roughton, the son of Thomas Roughton, of 111, Market Street, in whose possession it remains.

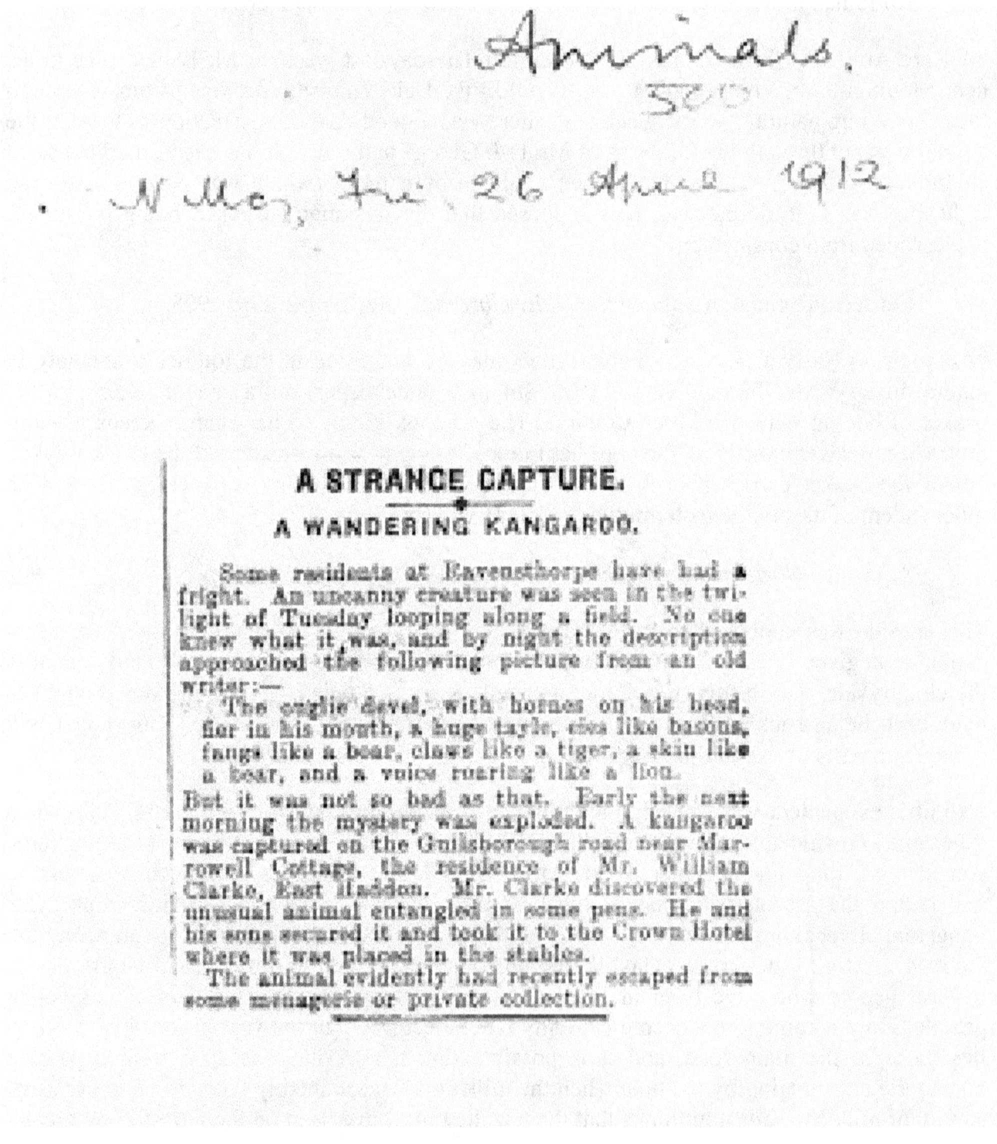

Animals.

500

N Mc, Fri. 26 April 1912

A STRANGE CAPTURE.

A WANDERING KANGAROO.

Some residents at Ravensthorpe have had a fright. An uncanny creature was seen in the twilight of Tuesday looping along a field. No one knew what it was, and by night the description approached the following picture from an old writer:—

The onglie devel, with hornes on his head, fier in his mouth, a huge tayle, eies like basons, fangs like a bear, claws like a tiger, a skin like a bear, and a voice rearing like a lion.

But it was not so bad as that. Early the next morning the mystery was exploded. A kangaroo was captured on the Guilsborough road near Marrowell Cottage, the residence of Mr. William Clarke, East Haddon. Mr. Clarke discovered the unusual animal entangled in some pens. He and his sons secured it and took it to the Crown Hotel where it was placed in the stables.

The animal evidently had recently escaped from some menagerie or private collection.

- A kangaroo. *Northampton Mercury*. April 26th 1912.

"Some residents at Ravensthorpe have had a fright. An uncanny creature was seen in the twilight of Tuesday (i.e. the 23rd) looping along a field. No one knew what it was, and by night the description approached the following picture from an old writer:- The ouglie devel, with hornes on his head, fier in his mouth, a huge tayle, eies like basons, fangs like a boar, claws like a tiger, a skin like a bear, and a voice roaring like a lion.

But it was not as bad as that. Early the next morning the mystery was exploded. A kangaroo was captured on the Guilsborough road near Marrowell Cottage, the residence of Mr William Clarke, East Haddon. Mr Clarke discovered the unusual animal entangled in some pens. He and his sons secured it and took it to the Crown Hotel where it was placed in the stables. The animal evidently had recently escaped from some menagerie or private collection".

- A Tasmanian cat (sic). *Northampton Mercury*. July 19th 1912.

Now this is very interesting. As you can see from the photograph this animal in question is not a cat, but what looks like a ring-tailed lemur. Nor are these lemurs native of Tasmania, but Madagascar.

So why does the text refer to "a Tasmanian cat"? Were there ever colonies of ring-tailed lemurs in Tasmania in the early 20th century or something similar and, if so, what became of them? Could a CFZ Australia member reading this look into it?

"The photograph is of a Tasmanian cat found on the line near Weedon, and given by the railway officials in charge of Mr B. Southgate, of the Horseshoe Inn, Weedon, until the owner can be found. Mr Southgate is making use of the opportunity by collecting for the Northampton Hospital".

- Siamese wild cat. *Northampton Daily Chronicle*. October 10th 1916.

Now here is another strange one. The piece doesn't seem to be referring to a domestic cat, but when I Googled "Siamese wild cat – images" I only came up with images of domestic cats or hybrids. Please can anyone tell me what "Siamese wild cat" would be referring to in 1916?

Furthermore, a cat, or most cats are bigger than stoats, (see below), so do we have here a cat similar in size to a stoat?

"The other night Mr Knight, a farmer of Chalveston, saw what he thought to be a stoat attacking one of his ducks. He shot and killed it, and then discovered the animal to be one of a species that he could not recognise, on taking to Mr H. H. Bryant, taxidermist, of Wellingborough, however, that gentleman was able to recognise it as a Siamese wild cat. How the animal, which, though small, is powerful and dangerous, got in this locality is unknown, but probably it had escaped from a private collection.

- Grey and red squirrels. *The Times* February 7th 1929.

This is the only item from a non provincial newspaper; nevertheless I include it here because I found it in the Northampton library and because, even though it is 80 years old, it is an early example, seemingly sincere of a certain belief about grey squirrels. This is only a very small extract:

"A few years ago my two elder sons saw a grey squirrel leave a red squirrel's nest in a tree close by our house, Iridge Place, near Etchingham, in Sussex. One climbed up and found in the nest a dead red squirrel, warm and bleeding."

Not only did I enjoy myself with these cuttings which now have a wider viewing nearly 12 years on, but Richard Freeman of the CFZ visited me and we had a drink at a town centre pub and went to a secondhand book shop where Richard found a book on animal mysteries. I bought *On Safari* by Armand Denis which has a photo of the remains of a four tusked elephant and a book called *Birth of Toads* by Elvig Hansen.

REFERENCE

(1) C. Fort. Book of the Damned in *The Complete Books of Charles Fort*. Dover Publications, Inc. New York. 1974. p92.

SUNDAY, FEBRUARY 15, 2009

GUEST BLOGGER RICHARD MUIRHEAD: A couple of interesting stories

I found the following two stories in the *Macclesfield Courier and Stockport Express* or *Cheshire General Advertiser*. (MCSE or CGA).

1. "Wild Man - A wildman was lately caught in the forests of Hungary, by a Wallachian, who first perceived him seated upon a tree, eating leaves. He appeared to be 24 years of age, and had his back and chest entirely overspread with a thick covering of hair: his skin was a dark yellow. He expressed only confused cries and a sort of murmur. He became melancholy at the sight of a forest or a garden and sought to hide himself in such places. After having been kept 2 or 3 years in captivity, he began to eat dressed meat, and from that moment his manners

became less savage; and his body losing a considerable portion of its hairy covering, became less yellow. He is now quite civilized, and performs the office of water-carrier; but never advances so far as to be able to speak words and sentences in succession.

2. An eagle of an immense size was shot lately at Heaton Norris near Stockport by Mr Geo. Bromiley of Gorton, which measured from tip to tip of the wings 7 feet 10 inches, in length 3 feet 4 inches, and weighed nearly 18 pounds. This mighty monarch of the air was fired at when in the very act of attacking a young pig, and being only slightly wounded in the shoulder and back part of the head great pains were taken to secure him alive, but from the very stout resistance he made, it became necessary to kill him to prevent his escaping. He is now in the museum of Mr Priestnall, surgeon, in Stockport, and is supposed to be the largest bird of the kind ever shot in England.

REFERENCES

1. MCSE or CGA March 23rd 1811
2. MCSE or CGA March 30th 1811

TUESDAY, FEBRUARY 17, 2009

MORE FROM MUIRHEAD....

Richard Muirhead has been down the library again.

As regular readers will know, I have been friends with him for 40 years now, since we were kids together in Hong Kong. He is undoubtedly one of the two best researchers I have ever met; he and Nigel Wright both have what Charlie Fort would have no doubt called a wild talent; a talent for going into a library, unearthing a stack of old newspapers, and coming back with some hitherto overlooked gem of arcane knowledge. Today he writes saying that, in his opinion, the following two new stories are not thoroughly unusual, but they do border on being strange. The first, the Scottish wolf, found dead on the east coast of Scotland, dates from April 1811, and the giant (ish) snake from August 1811.

"There was lately found at Tyringham, near Dunbar, the dead body of a large Wolf. There were several wounds on its head, and a cut on its neck, and from the appearance of the body it has long been dead. It was immediately skinned and stuffed, and is in good preservation. The colour is light, dusky yellow a black ridge down the back, and nearly white in the belly and breast. It has a sharp snout, erect ears, strong foreparts and a bushy tail. The length from the snout to the tip of the tail is 6 feet. The legs are shorter than usually described. It is conjectured the creature had been on board some of the vessels lately wrecked on the coast." *Macclesfield Courier* and *Stockport Express or*

Cheshire General Advertiser (MC & SECGA) April 27th 1811.

According to J. E. Harting the last wolf was killed in Scotland in 1743 [1] but M. Carwardine reports a date as late as 1848. [2]

Now here, the giant serpent of Florida:

> "A serpent of extraordinary size has been recently discovered in the Mississippi swamp, a few miles above the village. I have conversed with three gentlemen of unquestioned veracity who have seen it. They agree in their description, which is, in substance that the monster is in body considerably larger than the ordinary man, beautifully stripped with gold and green, rich beyond conception - the length is not accurately known, but is supposed to be from 15 to 23 (or 25, type unclear) feet. One of the gentlemen with whom I conversed shot at it with a rifle, when it emitted a very offensive smell, from which he supposed he had wounded it. However, as it was seen the succeeding day, it is presumable that if it was wounded, the wound was slight. This is the first serpent of such large size I have heard of." *(M.C. & SECGA) August 17th 1811.*

(1). J. E. Harting. *A Short History of The Wolf In Britain.* p.90
(2)   M. Carwardine. *The Guinness Book of Animal Records.* p.42

SATURDAY, FEBRUARY 21, 2009

MUIRHEAD'S MYSTERIES: A white hare in Devon and the remains of a huge bird off Siberia

I have today come across two new stories from the *Macclesfield Courier* and *Stockport Express or Cheshire General Advertiser.* (MC & SE or CGA) The first story is about a white hare from Devon, the second is far more controversial, an account of the remains of a huge bird found on an island in Arctic waters off the north coast of Siberia. Both stories are from 1811. A Google search for the name "Hedemstrotni" found nothing, but "New Siberia" does exist and it was discovered c.1806. Huge birds have been reported from North America but I was unaware of any from Siberia. Make of this what you will …

> " A white hare was killed last week at Puddington, Devon by the Rev. Mr Hole's harriers. A similar instance occurred about 50 years ago in the same parish." MC & SE or CGA October 26th 1811

"Hedemstrotni; the Russian naturalist, who recently examined the newly discovered Island called New Siberia in the Icy Ocean found on it three birds claws a yard in length; and the roving Jakute (?) related that they had sometimes found feathers, the barrels of which were capable of admitting a man's clenched fist. MC & SE or CGA. November 9th 1811.

Could "Jakute" now mean Yackut, i.e. of the Russian region now known by that title?

Little Weasel said: Before around 1800 a J was often used instead of a Y when writing, so it could quite possibly be a transliteration from the Cyrillic, that has used a J instead of a Y.

WEDNESDAY, FEBRUARY 25, 2009

MUIRHEAD'S MYSTERIES: A giant rat, an entombed bat and an imprisoned toad – 1812

My blog today covers the well-known Fortean phenomenon of entombed animals, but this time with a difference: Entombed toads will be familiar to the dedicated Fortean, but entombed bats may be less familiar.

There was a case of an imprisoned bat or bat-like creature in the Forest of Dean in Gloucestershire, U.K. in the 1920s or `30s, but in a fit of paranoia about the occult (because it seemed so weird) about 15 years ago I threw the evidence, in the form of copies of the newspaper cuttings, away. So part of this blog (see below) is an attempt to make up for that error.

Enjoy!

"A rat, of astonishing size, was lately killed at a public house in E. Clarendon, near Guildford [1] it measured from the tip of the nose to the end of the tail, 2 feet 3 inches, and was of proportionate bulk." *Macclesfield Courier and Stockport Express* or *Cheshire General Advertiser*. (MC & S.E or C.G.A.) March 21st 1812.p.3

"We are informed by a correspondent, that a bat was lately discovered at Astbury [2] in a solid block of stone quarried seven feet from the surface. The bat is described as having ears 7/8 (? figures unclear) of an inch in diameter and curvated like a ram`s horn; it was much larger than the common bat, and on exposure to the air seemed to languish. We shall be obliged to any naturalist amongst our numerous readers, who can inform us what species of bat can live in stone, or what kind of stone is sufficiently porous to permit animal respiration." *MC & S.E.* or *C.G.A.* March 21st 1812.p.3

"A live toad was found eight feet below the surface in a solid whitestone rock, in a quarry near Lochrutton.[3]" *MC & S.E.* or *C.G.A.* June 6th 1812. p.3

(1)   Surrey
(2)   Cheshire
(3)   Dumfries

MONDAY, MARCH 2, 2009

MUIRHEAD'S MYSTERIES: 600 snakes and a very big fish

Just a pair of short items today from the archives of the *Macclesfield Courier* in keeping with my custom of presenting pairs of cryptozoological or Fortean stories from the past, an aspect of cryptozoological research I hope to build upon in the months and years ahead.

Could this item below have been a seal or small whale? Not being a marine biologist, or even much of a zoologist, I am unaware of the normal weight of these creatures.

"An extraordinary fish was caught last week near Chertsey-bridge [1]: it has no scales upon it, is of a lead colour, and weighs 300lb." *Macclesfield Courier* July 25th 1812 p.3.

"On Tuesday last 600 snakes which had nestled in some old manure, lying on a field at Boltham, near Lincoln, were destroyed." *Macclesfield Courier* August 1st 1812. p.3.

Tragically this was the all too common fate of these "snake swarms" which turned up in huge numbers all at once like this in the 19th century.

(1)   In S.W. London probably

WEDNESDAY, MARCH 11, 2009

MUIRHEAD'S MYSTERIES: A Plymouth shark and a white swallow

After an absence of about a week due to the need to help with new work being done on my house, I have now collected a few more examples of early nineteenth century Fortean Zoology, from the archives of Macclesfield library, which I present here:

"A few days since, as one of the Lancashire Militia was bathing in Mill Bay, near Plymouth and was returning from a lugger to which he had swam, the spectators became alarmed by his repeated shrieks - A boat having been sent immediately to ascertain the cause of his cries, he informed the rowers that he had been followed and repeatedly bitten by a fish of enormous size, and on

examining his legs, they were found to have been bitten severely in several places."

It is probable that a shark may have followed the mackerel into Plymouth Sound, as the approach of this kind of fish to the western shores is generally marked by the appearance of fish of enormous size, which prey upon them. *Macclesfield Courier and Stockport Express* or *Cheshire General Advertiser. August 8th 1812. p.3*

And

.

"A white swallow was last week killed in Bulk, Lancaster, and is intended to be preserved for the inspection of the curious in natural history."

What HAPPENS to all these preserved Fortean specimens?!

TUESDAY, MARCH 17, 2009

MUIRHEAD'S MYSTERIES: Tritons and Grampuses

Here are two new stories, both of a marine nature. The first seems totally inexplicable...You must decide!

"A Triton - On the 31st of July {1812} an extraordinary animal was seen by five fisherman, in the creek of Port Mesin (?)(Murbian) Its shape resembled that of a man. It had arms, and the bust was completely human, but the lower part terminated in a fish`s tail. Its head was bald, with the exception of fore parts, on which was a bunch of black hair, and another bunch was perceptible upon the chin. The seafaring people who have sent us these particulars, had time to observe the monster at their leisure; it was within half a musket shot of the shore, between two boats, but they were afraid of it, and did not go any nearer"

(From a French Paper) *Macclesfield Courier*. September 5th 1812 p.2

"An enormous fish, believed to be a grampus, [1] was caught off Brighton...The havoc which this stupendous sea monster (which measures in length upwards of 31ft, and is nearly 16ft in girth) made in the herring nets that entangled it, cannot be repaired for a sum so moderate as £50, but already triple that amount has been made from its exhibition at 6d per head. The weight of the fish is computed to be from 5 to 6 tons; and the oil of it, if properly extracted, it is imagined will be worth from £100 to £150. "

Macclesfield Courier November 14th 1812.p.3

(1) Grampus is an alternate name for the orca, the white whale, and various species of porpoise.

1 COMMENT:

Max Blake said...
First one must be a dugong/manatee surely?

TUESDAY, MARCH 24, 2009

MUIRHEAD'S MYSTERIES: A giant basking shark and three white hares

Richard Muirhead is an old friend of the CFZ. I have been friends with him for 40 years now, since we were kids together in Hong Kong. He is undoubtedly one of the two best researchers I have ever met; he and Nigel Wright both have what Charlie Fort would have no doubt called a wild talent; a talent for going into a library, unearthing a stack of old newspapers, and coming back with some hitherto overlooked gem of arcane knowledge. Twice a week he wanders into the Macclesfield Public Library and comes out with enough material for a blog post...

I've been ill in bed most of the day with an appalling cough and cold but fortunately I have a small back log of animal stories from the pages of the *Macclesfield Courier* from late 1812 to mid-1813: Selling a basking shark for £600 seems a lot even by early 19th century standards:

"Another fish of the basking shark species, measuring in length 32ft, girth 19ft, caught off Brighton on Friday se`nnight* (sic), in the herring nets of an industrious fishermen, named Collins-sold next day by Dutch auction, for £600 to some person who mean to exhibit it in London." *Macclesfield Courier* December 12th 1812. p.3.

This would make it one of the largest ever basking sharks? And if it was not a basking shark, what was it?

"A few days ago three white hares were found in a field of Mr Maws (?) of Hacknew (?) near Scarbro` about one month old." *Macc. Courier* June 25th 1813.

Se'nnight = seven nights (one week), an archaic usage, though fortnight (14 nights, or two weeks) is still widely used.

THURSDAY, MARCH 26, 2009

MUIRHEAD'S MYSTERIES: Giant pike and a weird whale

For some reason the early years of the *Macclesfield Courier* were full of stories of somewhat unusual behaviour by aquatic creatures both inland and on the open waters within and around the British Isles. For example:

"A pike was lately caught in Windermere Lake of 30lbs weight: but a larger was once caught in the following extraordinary manner:- A calf belonging to a gentleman at Hawkshead was heard to make an uncommon noise by the side of the river, and on-going up to it, there was a large pike seen hanging from its nostrils, which it is supposed the fish had seized while the calf was drinking. The calf had dragged it about fifty yards from the river, and the pike was killed with a stone. It weighed 45lb". *Macclesfield Courier* July 24th 1813.p.2

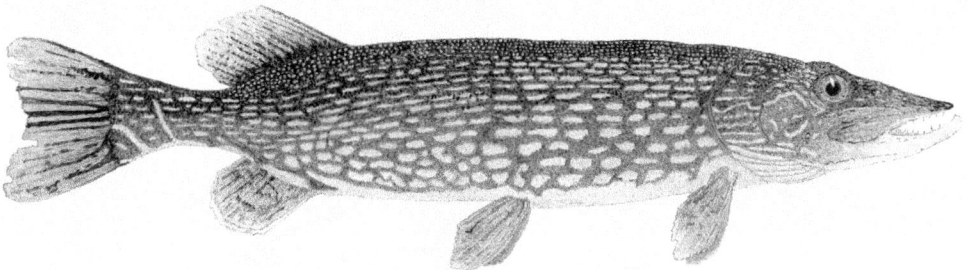

"A whale, nearly 30ft in length, was lately brought ashore in the neighbourhood of Irvine, in Scotland." *Macc. Courier* August 7th 1813.p.2

THURSDAY, MARCH 26, 2009

How the bloody hell does Muirhead find this stuff?

Dr. Stokes's red cross embrocation night and morning. These can be obtained at Mackey's Medical Hall, Newtownards.

THE SEA SERPENT

A REPORT FROM THE COPELANDS.

KILLED AND BEACHED.

Following the reports circulated recently as to the presence of a strange sea monster in Belfast Lough, a letter has been received from a resident on the Copeland Islands. He states that on the 5th inst great excitement was caused on the islands when it became known that a huge snake-like fish had been stranded in the shoal on Horse Point. This correspondent states that he and his brother were out walking, when they observed the water in the shoal being lashed about as if by a whale. The tide was out at the time, and on approaching the spot they were amazed to see a monster fish swimming about. Too terrified to go any closer, they were at a loss what to do; but at length, the correspondent, realising that the incoming tide would liberate the monster, despatched his brother for a gun, and told him to bring the boat round.

"It took us all our time," he states, "to kill the beast, and it was only after four shots had been fired into him that he stopped kicking. We then grappled him, but try as we might we could not get him to budge, so John went and brought two other men and a pony, and amongst us we beached him at last." Describing his capture, our correspondent states that he measured it and found it to be nearly 30 feet long, and about 6 feet round at its upper fins. The body tapers to about 6 inches at the tail, which is fan-like. There are three large fins, two on the back and one on the belly. The mouth, nose, and eyes resemble those of a conger eel, but are about five times as large. The body is covered with seales. The writer says that he is an old man, has lived all his life on the Copelands, and has seen most queer fishes, but never anything like this. He states that if any Belfast gentleman would care to examine the monster he or any of the residents on the island would on being signalled for take them from Donaghadee pier to where the body is beached. He adds that he would have communicated with us sooner but for the fact that during the past two days the weather has been too wild to permit of getting across to the mainland.

TUESDAY, MARCH 31, 2009

MUIRHEAD'S MYSTERIES: Suicidal dogs, and toads in an iron foundry...

Dear folks

For today's instalment of Muirhead's Mysteries I pass through the autumn of 1813 and encounter a break in the marine related features, only to uncover a case of suicidal dogs on the island of Samson in the Scilly Isles, (a strange coincidence here as the Biblically aware amongst you will remember that Samson did in fact commit suicide in the Philistine temple) and toads acting strangely:

"SINGULAR CIRCUMSTANCE: Last week, all the dogs in the Island of Samson (Scilly) in number about 14, ran simultaneously into the sea, and were drowned together! No cause whatever can be assigned by the inhabitants for this extraordinary occurrence. The dogs appeared perfectly well a short time before this event took place." *Macclesfield Courier* October 2nd 1813. p.2

"Remarkable circumstance- A few days ago, two living toads were found in the centre of the Cupola Furnace, at Mr Barnett's Iron Foundry in Skipton." October 16th 1813.p.3

I am familiar with the folklore of salamanders withstanding fire but not toads.

SUNDAY, APRIL 05, 2009

MUIRHEAD'S MYSTERIES: Feathered fish and eagle tales

In response to Jon's plea for more effort on the blog front, I have decided to submit three between now, (Sunday afternoon) and next Thursday evening, so here we go. In a strange coincidence, a few hours before I read the blog of a few days ago (sorry, I forgot to note down the exact date) about the furry fish, I noticed the following:

> (I had that morning over the phone asked to use the microwave
> instead of the microfilm. D'oh!)

"Mr Brougham, of the museum, Maryport [1] who got lately (?) into his possession a fish covered with feathers, has since caught one covered with hair." *Macclesfield Courier.* December 11th 1813 p.2

1.   There are two Maryports in the United Kingdom. One is in Dumfries and Galloway, Scotland the other in Cumbria on the coast. A Google search on "Maryport

Museum" uncovered a Maritime Museum in the latter Maryport, with a whales tooth etc. So I e-mailed them yesterday, Saturday April 4th, to ask them about the above mentioned fishes and I await, in hope, a reply.

"A fine eagle was killed on Friday se'night by the game-keeper of T. Thornhill Esq of Riddlesworth [2] In length from the end of the beak to the end of the tail, was three feet, the breadth, when its wings were extending, seven feet one inch, its weight nine pounds. *"Macclesfield Courier* January 15th 1814 p.2

2. Riddlesworth is in Norfolk

THURSDAY, APRIL 09, 2009

MUIRHEAD'S MYSTERIES: Tortoises in Essex, and a letter from Ken Livingstone

Dear folks,

It's time for my update from the improbable world of Richard Muirhead. This evening I am giving the *Macclesfield Courier* a brief skip (we will resume with it after Easter) and turning to a letter I had written on May 7th 1997 when I was living as a student in Oxford. (Not at Oxford University I hasten to add, but at Oxford Brookes University where I was doing my M.A) I also publish for the first time a very brief and absolutely genuine letter to me about newts from Ken Livingstone M.P. as he then was in response to mine to him asking-was the "effet" a newt, or not?

> Dear Mr Muirhead
>
> By luck I saw your letter about tortoises in the wild in a recent issue of the *Cambridge Evening News*. I hope the following information helps you.
>
> Between 1959 and 1964 my uncle kept a caravan for weekend and holiday use at Great Gibcracks Farm, near Chelmsford in Essex. The farm was about half a mile from the road, up a long and rutted drive. It was an extraordinary place, an Edwardian model farm built by an eccentric who placed busts of Dante and other poets in copses had a swimming pool dug and planted exotic trees and shrubs. By the 1960s it was falling to pieces about the current owner.
>
> Beyond the farm were two cottages and then woodland. The nearest houses to the farm, apart from the cottages, were almost a mile away.
>
> In late May or early June 1964 my younger brother, Nigel, and I found a tortoise moving along a furrow in a ploughed field immediately alongside the farm. (The field had been ploughed at Easter then left.) I can't be more precise about the date; I do know that I'd been given a weekend

away from preparing for the O levels looming over me.

The tortoise just fitted on my palm so would have been almost 6" long. We had kept several tortoises as pets in the past; this one had a darker shell than I could recall seeing before and yellow mottling between the eyes. It appeared to be a spur thighed tortoise.

We went to the farm, the cottages and then to the houses nearest the field but no one knew anything about the tortoise. Using the local bush telegraph we let people living farther away know but the tortoise was never claimed and became our pet, with the name Shostakovich-why that name was selected I can`t remember. Shostakovich was found dead in summer, 1967.

In 1980 Nigel mentioned he had talked with a man who lived in a village called Bicknacre in the 1950s and 1960s; Bicknacre is not far from Gt. Gibcracks and we sometimes walked there through the woods. The man said that he twice found tortoises in the fields but had never traced their owners. Sadly, Nigel died in 1990 and I remember no details, like the man`s name.

I hope that your research is fruitful; the adjustment of animals to British conditions has always interested media have wondered if the Surrey puma and Exmoor beast are abandoned pets that are now acclimatized. I am sure you know about the scorpions that colonized the railways station at Ongar in Essex.

It would be more interesting though to discover native tortoises and native pumas. I hope one day to read about your research.

Yours sincerely.
J.S. Holford-Miettinen

I rang a Cambridge number about 2 months ago to try and find out more about the tortoises but got no reply. Now for "Red Ken`s" letter, it is very brief:

April 22nd 1996

"Dear Richard Muirhead, Thank you for your letter of 15 April about the terms eft or evvet. They were just other names for newts-not crocodiles or anything else.

Yours sincerely,
Ken Livingstone.

I have a lot of respect for a person as busy as he must have been to reply like this, albeit briefly. After Easter I will continue with the *Macclesfield Courier* and octopus invasions of the south coast of Britain, an historical approach. Bye for now and Happy Easter! Rich.

TUESDAY, APRIL 14, 2009

MUIRHEAD'S MYSTERIES: White starlings and a cure for vipers

Now that Easter is over I`m returning with Fortean zoology from the *Macclesfield Courier*.

Now we`re up to January 1814:

"RARA AVIS-On Tuesday se`enight a person of Boston [1] shot a white starling, which was one of a flock of birds of the ordinary colour, flying in the pastures round that town. Being only slightly wounded, it is yet alive. The gunner observed another white starling in the flock."
[1] Boston is, by the way, the town in Lincolnshire
Macclesfield Courier. January 22nd 1814.p.2

"Prof. Mangeli (?) has published in the *Milan Journal*, a long report upon the action of the venom of vipers. He states, as the result of his experience, the ammonite is the only sovereign remedy for the bite of those reptiles, and that opium and musk, which have been hitherto prescribed, have no certain effect".

THURSDAY, APRIL 16, 2009

MUIRHEAD'S MYSTERIES: Golden maids all in a row

Dear folks.

First I must apologise: I was going to present extracts from my newspaper archive of reports of octopi strandings along the south coast of Britain from 1937-1977 but lack of time means I am unable to do so today. However I will be at CFZ Headquarters in Devon early next week so if there is enough time then and it is O.K. with Jon I will present the information from there. Meanwhile, I have a report from the winter of 1814 of SOMETHING like an octopus or jellyfish off Brighton:

"During the last few days, a great number of the fish called Golden Maids [1] were picked up at Brighton beach, and sold at good prices. They float on shore quite blind, a state to which they are reduced by the snow; and it is a fact well known, that after heavy falls, these fish are always thus found in great abundance…"

And, wonders never cease….

"On the morning of the 18th ult the frost was so severe, that a wood pigeon was taken alive in Pilfirrane-garden, near Dunfermline, having its feet frozen to a cabbage, on which it had alighted but a few seconds before!" Both stories: *Macclesfield Courier.* February 12th 1814.page 2.

[1] Golden maids: a cursory glance at Google couldn`t reveal what `Golden Maids` were or are.

THURSDAY, APRIL 30, 2009

MUIRHEAD'S MYSTERIES: Octopus invasions (and more)

Dear folks,

Sorry I`ve been away for so long, I have no excuse so I won`t attempt to make one! I am now presenting part 1 of an archive of newspaper cuttings relating to invasions of octopi along the south coast of Britain in the early 1950s. Some of the dates are illegible as are some of the names of the newspapers (but probably the forerunner to the *Brighton and Hove Leader*) so I have made educated guesses. I have three reports from the 1970s. Not every report is included, only those deemed to be of interest, and at the same time not the whole report.

August 16th 1950: "They`re a Catching Complaint": Instead of the usual dabs and whiting, fishermen at the end of Brighton`s Palace Pier have been pushing ashore octopuses. On Monday no fewer than were caught were caught……Pier master Capt. Fred Weeks told a reporter, "We are used to catching one or two each season, but this has been a most unusual crop. I`ve been on this Pier since 1928 and I`ve never known so many to be caught together…..

ODD CATCH

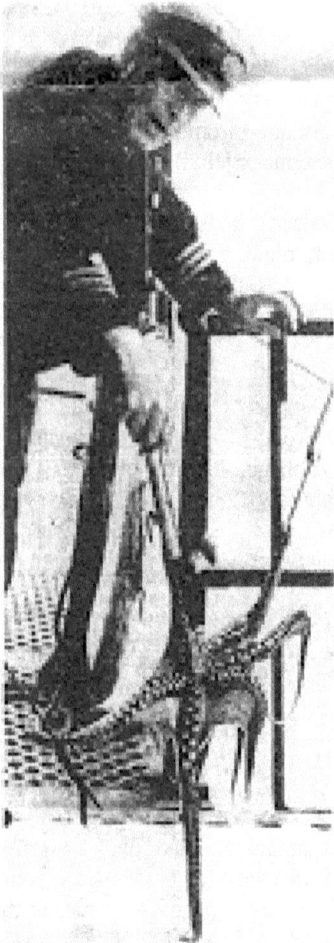

One of several octopuses which anglers fishing from the Brighton Palace Pier have landed during the past few days. This 4ft. octopus was caught last night and is being examined by Capt F. Weeks, the Piermaster, before being thrown back in the sea.

FOOTNOTE: Largest octopus "plague" was in 1899. It had such a ruinous effect on the shell fisheries that lobster fishermen were forced to seek other employment. A more minor "plague" occurred at Brighton in December 1922, when the beaches were littered with thousands of dead octopuses thrown up after a storm."

August 28th 1950: "Kill The Octopus" Other unwelcome marine "monsters" not usually frequenting our shores, have been joining in the cross-Channel swim, which again seems to prove that the Channel is warming up. B.W. Downes Castle-square Brighton"

September 8th 1950: "Octopuses (?) Lobster Pots Waters Infested". Nothing much new on this date. "Sisley is not the only place where the octopus plague is being experienced. It is affecting fishing on the Dorset and Devonshire coasts, and some hundreds of small ones have been washed up on the beach at Brighton."

September 14th 1950: "Octopus Menace." Not much of interest here, a final comment stating:

"Fishermen hope, however, that nature will solve the problem in the same way that she set it. The advent of colder weather this autumn will either kill the octopuses or drive them to warmer grounds."

September 15th 1950 [*Hants Post*. No headline.] ….. "It is believed that the octopuses get too crowded on the French side of the Channel where they breed and come over to the coast of Southern England…."

September 16th 1950: "Sussex Waters Plagued by Octopuses" This month hundreds of these loathsome molluscs some measuring over three feet across, have been reported near Selsey where they are causing havoc in the shell-fishing industry….[here the journalist goes on to describe giant octopi:…. "The boats of native Japanese fishermen have been upset by such huge creatures…"]

October 30th 1950: Octopus activity was also present off Worthing: "Octopus Caught" A baby octopus, about 3ft 6ins long was caught by Mr W. Belton and Mr E. Edwards in their fishing nets off Worthing early today. "It was hanging on to a red mullet", said Mr Belton.

Here concludes part 1 of the review. I will attempt to conclude with Part 2 before I go abroad on May 9th.

Now a look at a curious observation by the 17th century naturalist and antiquarian John Aubrey in MS Aubrey 1 in the Bodleian Library, Oxford: "I have been told heretofore, that in the ruins of Bampton Castle Oxfordshire have been found Scorpions....let it be further examined."(c.1685) also, "In Warrens (?) are found rarely some stotes, quite white, that is they are ermines. My keeper of Vernditch Warren * hath showed two or three of them to me. Every Warrener knows this to be so but all stotes are white under the bellies". [MS Aubrey 1.* Wiltshire-Dorset border?]

FRIDAY, MAY 22, 2009

MUIRHEAD'S MYSTERIES: THE BIG THREE:
Richard Muirhead

A few weeks ago we asked various bloggo regulars to tell us what their top three favourite mystery animals were ... And why?

I guess my favourite mystery animal, (and I place that deliberately in the singular, as it may have been one of the last of its kind) is the Giant Flying Snake of Namibia, or south-east Namibia's Keetmanshoop region of karst-like hillocks and caves, to be precise.

I first "discovered" mention of this cryptid in 1995 or 1996 in the basement of a second hand book shop in Charing Cross Road in London , where I picked up `These Wonders To Behold` Lawrence Green, (Howard Timmins, Cape Town, 1959) but it is mentioned in my essay on this cryptid in the *1996 CFZ Yearbook*. I am pretty sure that, apart from Karl Shuker`s writings and cryptozoological encyclopaedia, the Namibian Flying Snake is only mentioned briefly in *Searching for Hidden Animals,* by Roy Mackal. (Cadogan Books, London, 1980) But note, this is only as far as I am aware.

The interesting thing about this Flying Snake is its resemblances to a dragon in several respects, e.g. its habitation, on hillocks, its ability to fly, or rather, glide, its tendency to leave a noxious odour or slime. It also seems to straddle the boundary between "paranormal" and "flesh and blood" entities, rather like Sasquatch turning up in unlikely places like the Bible-Belt in the U.S.A. where their behaviour is very un-apelike, apparently.

Other interesting features of the Namibian Flying Snake is the light on its head, which hovers above the ground, rather like a U.F.O. is said to hover above the ground with a light, in the developed world. Finally, just what was Courtenay-Latimer's role in photographing the cryptids tracks after it attacked the young farmer`s son near Keetmanshoop in c.1942?

The living sabre-tooth tiger or *tigre de montaigne* and its more aquatic relatives is another favourite of mine. This is because I find it pretty amazing, a real prehistoric survivor if there ever was one. Unfortunately, as far as I know, there is very slender evidence for its existence, at least from Western sources. Who knows, perhaps in some dusty Chadian or French library, there are old newspaper cuttings? Or perhaps a genius with Google Earth will someday be able to track down its lair?

I first came across mention of the *tigre de montaigne* in Karl Shuker`s *Mystery Cats of the World* and was immediately hooked. Little is known about it, except its fierce roar and its banded red and white coat. Chad is a dangerous place, with warring factions and tension along the border with Sudan to the east. It is said to inhabit caves in the mountainous northern region along the border with Libya and near the central town of Mongo. I hope one day to look for it. Meanwhile those interested can check out my web sites.

The tatzelwurm is a somewhat sinister bipedal lizard or salamander-like animal said to inhabit mountainous regions of Switzerland, Austria and southern Germany. No definite photograph of the pale cryptid exist, but interestingly, the testimony of Maurice Masse outside the Natural History Museum in Paris in 1934 as to the appearance of the tatzelwurm, (as reported by Roger Hutchings in Alpine Enigma, *Animals & Men* 2 pp. 20-21) particularly that they only have forelegs, is borne out by documents in French such as `Un Animal Inconnu Dans Les Alpes` by Willy Ley in *La Nature*, Paris, June 1st 1938 pp. 366-367 and also in French `Le Tatzelwurm` by Georges Hardy in *Tribune de Geneve* November 28th 1969, `Portrait-Robot` by the same author February 6th 1970 and finally *Cherchez le Tatzelwurm*! Also by G. Hardy March 13th 1970.

The second of these articles includes a drawing of a tatzelwurm which shows a tadpole-like creature with beady eyes and only two forelegs! Unfortunately the quality of the drawings is too poor to reproduce here, but I will give a translation from the French of the part of the article which was headlined: `Schema le plus simplife de la forme de l`animal` or `The simplistic plan of the outline of the animal`.

An approximate form and colour of the body and the eyes, as I got to see them unexpectedly on the road, from the top of the village of Mordes, which goes to Forts de Dailly, to the right, an ancient walled fort, under the bushes. Not far from there are damp, humid caves and I think that the animal itself may have emerged from one of the caves (?), crossing the road, which was sunny this mid-September day, but cool, to go to the other side of the road, where, if I remember clearly, vegetable gardens and grass/lawns (were?) It was a beautiful day in Lavey-les-bains (hot) quite cool in Mordes alt.1,160m. The Tatzelwurm was on the road, dry but cool, three quarters across the road, on its walk, towards the orchard (?) Maybe he likes vegetables? As I told you during the interview I felt so disgusted by this beast that I did not stay in its neighbourhood. The colour was pinky white, like a foetus (i.e. a juvenile? - R. Muirhead`s note). Staring, round, protruding eyes surrounded by a pronounced black circle gave to the animal a provoking and threateningly ugly appearance.

The article gives two drawings of the Tatzelwurm, one pale and one dark. This article was translated from the French into English by Ronan Pellen in Cork in the mid-1990s, who insisted he was a Breton, not French.

FRIDAY, MAY 29, 2009

Richard Muirhead: Reed wolves and swarming blues

Richard Muirhead is an old friend of the CFZ. I have been friends with him for 40 years now, since we were kids together in Hong Kong. He is undoubtedly one of the two best researchers I have ever met; he and Nigel Wright both have what Charlie Fort would have no doubt called a wild talent; a talent for going into a library, unearthing a stack of old newspapers, and coming back with some hitherto overlooked gem of arcane knowledge. Twice a week he wanders into the Macclesfield Public Library and comes out with enough material for a blog post.

Every year the season at Royston became more and more exciting, aberrations were very numerous and varied in form and everyone on the Heath, even the most unlucky collectors, picked up something outstanding. But the end came dramatically. During the summer of 1918 you could see these Blues in tens of thousands. This was especially noticeable in the evenings, when they always assembled in sheltered dells and crowded on to tall grasses; you could stand in one spot and, without moving, count roughly five hundred of these little butterflies at rest, four or five often clinging to a single blade of grass. The following season collectors came to Royston as usual and were amazed to find that the numbers of Chalk-hill Blues were down to countable hundreds, spread over the whole area; well can I remember my own acute disappointment.

Sometimes it is difficult to give an exact explanation of any unusual event in nature but it seems in this case that a species of Ichneumon fly which preys on these Blues could be blamed for their sudden disappearance.

However, as you may have guessed his natural habitat is the library. This past few weeks, for the first time since I have known him, he has been out in the field doing research. He has been working with a butterfly conservation organisation in Hungary, and whilst he was out there he has carried out a few cryptozoological tasks for us.

He telephoned the other day, excitedly telling me about a swarm of blue butterflies he had seen; well over a thousand silver studded blues *(Plebeius argus)*. Such swarms used to be common in Britain, (as you can see from the second extract this week from L Hugh Newman's seminal *Living with Butterflies* (1967) which is posted on the left.

However, to the best of my knowledge they have not been seen in Britain for many years, so Richard was unsurprisingly excited.

But it wasn't just butterflies that attracted his attention. About ten minutes ago he telephoned me from a train just outside Brussels - he is on his way home, and he said that he had some interesting news about the reed wolf, about which I first read in Karl Shuker's *Extraordinary Animals Revisited.* Karl wrote:

"The Hungarian reed wolf was a small, mysterious form of wild dog existing in Hungary and eastern Austria until the early 1900s. In 1856, M. Mojsisovics named it *Canis lupus minor*, treating it as a small wolf, but the precise nature of its identity remained a debated issue long after that. In the late 1950s, this extinct enigma inspired a series of interchanges in various journals between Hungarian researchers Drs Eugen Nagy and János Szunyoghy. Nagy staunchly supported Mojsisovics's reed wolf classification, but Szunyoghy categorized it as a larger-than-normal version of the common jackal (in 1938, Dr Gyula Éhik had actually renamed it *C. aureus hungaricus*). However, the detailed studies of Prof. Eduard-Paul Tratz with the handful of museum specimens of reed wolf in existence provided persuasive evidence for believing that it had been an unusually diminutive race of wolf after all, an identity that has since won widespread acceptance."

It was difficult to make out what Richard was saying. There were train noises, crackling sounds, and he was on a mobile 'phone three countries away, and I am going deaf. But it appears that the controversy continues, over whether it was jackal or wolf, and also that the use of the past tense is not really justified. For, as far as I can gather, sightings continue to the present day.

More news when I get it (and BTW, Richard asked me to tell you all that *Muirhead's Mysteries* will be back next week)

Totally coincidentally, our new buddy Scotty Westfall over on the *Wildlife Mysteries* blog which gets better and better each day, has just published an article on the reed wolf which can be found at the link below:

http://wildlifemysteries.wordpress.com/2009/05/26/what-is-the-reed-wolf/

TUESDAY, JUNE 02, 2009

MUIRHEAD'S MYSTERIES: Back from Hungary

Dear friends,

As some of you may know, I have been in the Aggtalek National Park Region (see http://www.anp.hu/) of northern Hungary recording butterflies, in particular the migration and interaction between 2 species of Fritilaries, the *Melitaea telona* (pictured above) and the *Melitaea phoebe* or Knapweed Fritillary. There were other Fritillaries there and altogether by the end of the 3 weeks we had recorded about 60 different species of butterfly. We also saw or heard many moths, reptiles and amphibians, bats, birds and mammals. I will not spend any more time writing about this trip here, because I am writing an essay for a future edition of the C.F.Z`s *Amateur Naturalist* Magazine mainly concentrating on swarms and migrations of butterflies and I have also today posted about an hour`s worth of video footage to Jon which he will have to heavily edit if it is to be suitable for C.F.Z. TV! I will be putting all of my best digital photos onto a disc. Whist in Hungary we saw a few ragged Painted Ladies flying up a wooded slope on migration. This was on May 25th. One of the last butterflies we spotted was the rare, (to Hungary) Poplar Admiral *Limentis populi*.

If anyone has any information on why the Black Veined White (*Aporia crataegi*) disappeared from Kent in 1906 please can they inform Jon or I? Thanks.

Finally, as far as Hungary is concerned, I found this Latin inscription beneath a painting of a saintly character in a church near our hostel: So putting on my Dan Brown hat I managed to translate some of it:

Passus Sub Pontio Pilato Crucifixus Mortus Et Sepultus Docuit In Scithia Cruci Afixus AC 69. = Having Suffered under Pontius Pilate Crucifixus (?) Mortus= Died (?) Et = And Buried Docuit (?) in Scithia = Southern Russia Cruci Afixus (?)

This cannot refer to Jesus Christ as He died in 30 or 33AD. I may be able to provide a photo of the inscription in a few weeks.

This is all I have managed. Can anyone help? I would be most grateful. I met a bloke called Tim who was on the trip with his girlfriend. He told me about a "sea-serpent" washed ashore in 1996 on the Isle of Harris in the Outer Hebrides. It was so smelly and slimy that the locals dragged it to a moor and buried it. Its head was 8ft long and it was transparent and body grey and slimy. It was on the north coast at the next bay to Luskentyr beach.

Tim also told me of folklore of ravens in the Hebrides who fly to your door and knock on it. If you open the door the raven turns into a beautiful woman, comes in and you are never seen again. The final bit of cryptozoology refers to huge footprints seen some time in the winter of 2008 by Tim. The footprints were in wood land on the Derbyshire-Staffs border. They measured 7 inches long by 6 inches wide and 3 cm deep with claws.

TUESDAY, JUNE 09, 2009

MUIRHEAD'S MYSTERIES: The Herring Hog

Dear folks

In one of those strange coincidences, as I was planning today's blog, Jon drew my attention to Mike Hallowell's contribution of June 5th which mentioned The Great Herring Catch of September 1807 in his part of the world.

The coincidence lies in the fact that a few weeks ago whilst in one of my favourites haunts (other than the local karaoke bar!) the Newspaper Library in north London I came across a story which was indexed as being about `The Herring hog`. This was in the `*Dumfries and Galloway Saturday*

PORPOISE.—Page 537.

Standard of October 30th 1909.

It was doubly rewarding, I thought, "herring hog, what's this, a new type of cryptid?" and also, my ancestors were Scottish. But alas, on doing a Google search on `herring hog` I found out that this is in reality the common porpoise of the Atlantic and Pacific[1] So, it`s always best to check when coming across a possible new cryptid.

REFERENCE

1. http://www.thefreedictionary.com/herring+hog

THURSDAY, JUNE 25, 2009

MUIRHEAD'S MYSTERIES: Macclesfield wallabies are back?

Dear folks,

I have some new information on the wallabies said to roam the moors above Macclesfield. The *Macclesfield Express* has twice in the last few weeks reported on the status of wallabies supposed to have become extinct four years ago.

Firstly, on June 10th the cheesy headline *'Can Roo believe it?'* reported: "After years of speculation, it appears there ARE wallabies living in Swythamley." The rest of the article says nothing significant to add to this but does describe the animals as being in the plural.

Later, in the *Macclesfield Express* for June 24th, there was more substantial evidence of wallabies in this part of the north-west, with some interesting historical information. I quote: "Do these new photographs prove that wallabies are still living up in the hills near Macclesfield? Hiker Andy Burton said he captured one of the elusive creatures on camera while trekking up the Roaches with friends...Wallabies were introduced to the Peak District in the 1930s by the Brocklehurst family, when five of the animals escaped from their private zoo at Roaches Hall. It's believed up to 50 were living on the moors at one time, but many were hunted and fears arose they had completely died out...A spokeswoman for Peak District

Rangers service said there had been no official or confirmed wallaby sightings in four years."

TUESDAY, JUNE 30, 2009

MUIRHEAD'S MYSTERIES: Unusual stories from the English countryside 1750-1850

Dear folks

These are stories from an uncorrected proof of *News from the English Countryside 1750-1850* by Clifford Morsley (1979), which I found in a bookshop somewhere; I can't recollect where.

Boy eats cat
Cambridge. On Tuesday evening a country lad, about 16, for a trifling wager, ate, at a public house in this town, a leg of mutton which weighed near eight pounds, besides a large quantity of bread, carrots, &c. The next night the cormorant devoured a whole cat smothered with onions.

Cambridge Chronicle quoted in *The British Chronicle* 13 September 1770

Strangest Phenomenon within Living Memory
Birbeck Fell, September 23. The following circumstance, however improbable, may be depended upon as a matter of fact. A farmer's wife, in this neighbourhood, who attended duly to the milking of her cows morning and evening, observed for two or three mornings successively that her best cow was deficient in her usual quantities of milk; this made her suspect that some of her neighbours were not over honest, and communicating her suspicions to her husband, they resolved to watch all the succeeding night, which they did without making any discovery....

Following her thither they observed a most enormous over-grown adder, or hag worm, crawl out of the root of the bush, and winding up one of the cow's hind legs, apply its mouth to one of the paps,

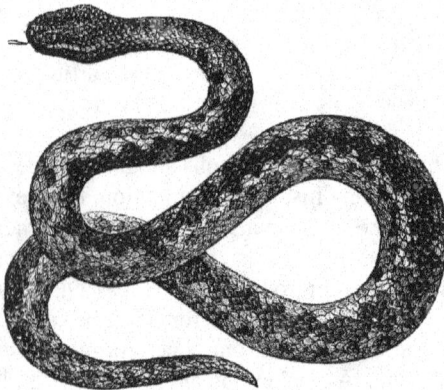

and begin to suck, which she suffered it patiently to do, till the farmer attacked it with a cudgel, and ere it could recover its den, kill it. It measured upwards of four feet in length and its skin, stuffed, may be seen at the farmer's house.

The whole is looked upon as the strangest phenomenon that has been known within the memory of the oldest man living. *The British Chronicle* 15 October 1770

Gigantic Gooseberry
A gooseberry was gathered in the garden of Thomas Tebbit, a gardener of Soham, at the beginning of August, which measured 4 ½ inches in circumference. *The Cambridge Chronicle and*

Journal. 27 August 1813

Rider in the Sky
The following story has appeared in several papers:

Some months ago a very singular appearance presented itself in the sky to several persons at Hartford bridge, near Basingstoke. About noon was distinctly seen by many persons, without any difference among them as to the form of the figures in the clouds, a man on horseback riding at full speed, pursued by an eagle, which soon darted upon his head, when he lost hold of the reins, fell backward, and eagle, horse and man were seen no more. The figures were apparently of natural size. *The County Chronicle* 10 February 1818.

Curious Fact
Mr Charles Parker, of Arundel, brought home three very young rabbits, which for the sake of warmth were placed before the fire. The house cat had kittened the same day, and on discovering the young rabbits showed great affection for them; on the following morning all the kittens but one were destroyed, and the rabbits placed under the care of the cat, who has ever since showed the greatest solicitude for their welfare, and they are now thriving under the kind offices of their feline foster-mother. *The Sussex Advertiser* 4 April 1831

Below: A similar incident from 2011

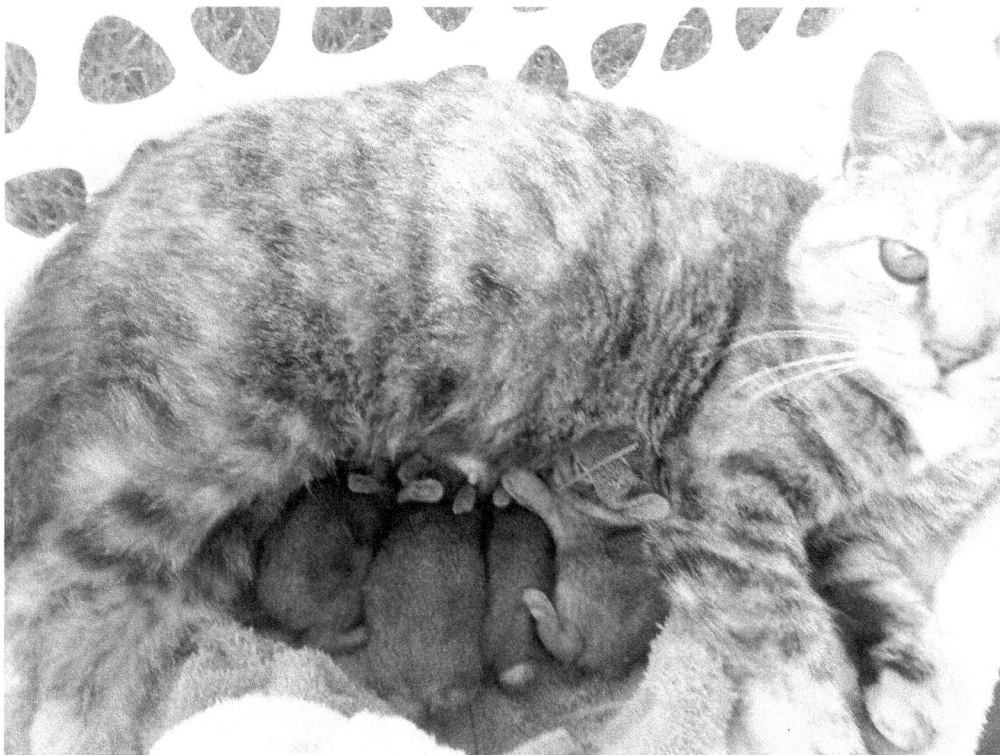

Another strange Hong Kong animal story from Richard Muirhead

7.5.1925

OVERLAND CHINA MAIL
23

WEIRD MONSTER.

"MOUTH AS BIG AS A BASIN."

NEW TERRITORY SCARE.

Farmer Who Was Swallowed Alive!

Reading like a challenge to zoologists is a story of a New Territory ferocious animal which devours human beings alive and has even, by its appearance, actually frightened people to death.

With a general appearance like that of a cow, a mouth "as big as a basin" and bristles like those of hedgehogs—such is the description of the strange animal in to-day's "China News," a local vernacular paper.

Lau Kwong, a Kowloon farmer is, says the vernacular paper, the unfortunate individual who encountered this ferocious animal and met his death.

Translated broadly, the story reads:—

In a glen near Taipo, Lau was chopping firewood when he suddenly saw this huge animal. Taking fright immediately, Lau ran away to return several days later.

Moving at a very rapid pace, the monster's gait, according to the version, was similar to that of a kangaroo as it waved its arms about and danced with its feet.

Lau took to flight with the animal in chase close behind as it had sighted him. Soon the farmer was overtaken and swallowed alive by the ferocious animal!

In the course of the last few days, a number of persons are said to have met their death through fear of the animal while others are in constant dread at even the mention of it.

The mainland has been the hunting ground of various kinds of wild animals, real and imaginary.

Several years ago, a full sized tiger was shot after causing the death of a European police sergeant by name of Goucher. Since then and before, tiger stories have been too numerous to be recorded. Spoor have been seen from time to time.

This latest story sems to have emanated from native police officers coming in from out-stations.

Not so long ago there was another mysterious animal. Some said that it was a bear, others a kangaroo.

Local sportsmen out for week-ends have bagged deer, aided by beaters and wonks and a few speak of having seen stray foxes.

Although this is the close season, the "China News" expresses the hope that adventurous sportsmen will rid the Colony of this mysterious wild animal and establish its identity.

WEDNESDAY, JULY 15, 2009

RICHARD MUIRHEAD: Polecats in Wiltshire

RICHARD MUIRHEAD WRITES: Hi folks, I found this article and blurred photo of a polecat today in my files. It is from *The Salisbury Journal* of June 14th 1984. I believe in the mid-1980s polecats had been absent from Wiltshire for something like 70 years.

If so, what was this one doing in Salisbury in 1984? The references to other sightings in the article are intriguing.

THURSDAY, JULY 30, 2009

RICHARD MUIRHEAD: More vintage Fortean zoology

Hello again folks; hope you`re all well!

I am presenting some cryptozoology and Forteana from various British natural history magazines from the period 1905-1912, which I gleaned whilst an MA student in Oxford in the mid-1990s. A few of these may have been mentioned elsewhere but they have never been gathered together all in one place. They should provide a safe reference tool for anyone with the time to research and a decent library nearby. *Country-side Ju*ne 17th 1905 p.89 African Viper, Denmark Hill (London?)

- Brusher Mills. (A snake catcher in the New Forest) August 5th 1905. p.199
- Odd coloured frog, N. Wales. December 9th 1905 p.55. "The frog described as having a back of a bright red colour, spotted with orange, and a pure white breast seen swimming in a brook in N. Wales, was an interesting colour variation of the common frog". Same page- bright yellow frog Hull.
- July 21st 1906 p.160. Acclimatisation of River tortoise in Thames valley and elsewhere in England- carnivorous-dark shell with bright yellow dots.
- September 1st 1906 p.230 A 2ft long grass snake killed in Ballymena.
- October 13th 1906 p.298 A grey slow loris from China at London Zoo [1]
- October 27th 1906 p.324 . "Sea creature off Scotland"
- June 15th 1907 p.83 Cat headed sea snake
- July 4th 1908 p.79 Another red British frog.
- May 15th 1909 p.341 A little red viper
- September 4th 1909.p.243 A little red viper in S. Wales
- This is my favourite one: October 9th 1909 p.326 A type of African water snake with an "electrical battery" (sic)
- *Country-side monthly* Exact date unknown. Vol 3 p.136 Magical cure from snake bite
- December 18th 1909 p.71 Spanish terrapin (in UK) "uttering faint but distinct "mews" "like a young kitten. I think the sound is produced by the drawing in of the breath....When alarmed he always hisses, but it is only lately he has taken to mewing."
- A similar story appeared in *Country-side* in 1912 p.410.
- Green lizard in Lancaster 1912 p.410

Jumping forward to – (Radio 4) April 28th 1991 "Bees this year are buzzing a semi tone higher than last year".

N.B *Country-side* and *Country-side monthly* are different magazines.

(1) It would be interesting to know exactly from what part of China this grey slow loris came from because around this time a slow loris was found tied to a lamp post in Hong Kong. Has anyone heard anything about this?

That`s all folks....

SUNDAY, AUGUST 02, 2009

New from CFZ Press

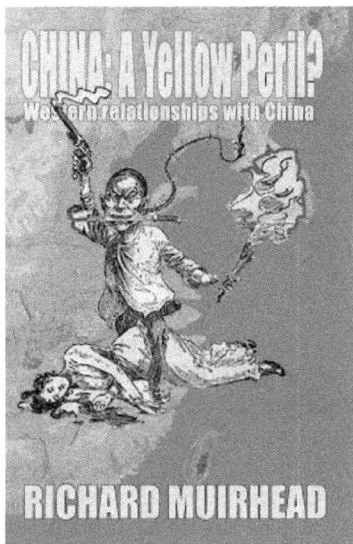

If some Victorian antiquarians are to be believed, contact between the Chinese Empire, and other Middle Eastern and Western Empires goes back to long before the birth of Christ; such as the ancient Egyptians and the Roman Empire. A Roman coin from the time of Hadrian in the second century of the Christian era was found in Oshkosh in Wisconsin in 1883, thought at the time to have been carried there across the Bering Straits to Wisconsin by way of Alaska by a Chinese person. Muirhead`s book *China: A Yellow Peril? Western Relationships with the Chinese* looks at a time period long after these very early contacts, to the beginning of trading links between the West and China in the Seventeenth Century, with the arrival of the Jesuit intellectual and religious leaders.

The impact of these individuals as well as the British, French, Russians, Japanese, Germans and Americans in the following three hundred or so years created a tension that resulted negatively in the West and elsewhere in the racist Yellow Peril scare; and positively in developments such as an appreciation of China as a cultured civilisation with trade in Chinoiserie and food stuffs. In fact, between the late eighteenth century and the early decades of the twentieth century there was a debate between detractors and supporters of China as either barbarian or civilised, with relationships between British and Chinese in the colony of Hong Kong perhaps surprisingly surviving the complex change of events in China that led to the rise of communism in rural and urban China from the 1920s onwards.

The Yellow Peril scare, essentially a fear of Chinese expansionism and morals, is the main subject matter of this book. Muirhead concludes that with the pressures brought upon the world by China`s massive economic growth and pollution comes the risk of a revival of the Yellow Peril scare."

Muirhead hopes his book will dispel a tendency amongst some commentators to portray everything in black and white and an unnecessary overwhelming guilt for colonialism. In fact, there were good imperialists and Victorians, and patriotic Chinese communists.

http://www.amazon.co.uk/China-Yellow-Peril-Richard-Muirhead/dp/1905723431/ref=sr_1_1?ie=UTF8&qid=1249208101&sr=8-1

RICHARD MUIRHEAD: Ropen evidence

Hello again. In this blog I present what might be new information on the mystery flying creature of Papua New Guinea, the ropen, which some believe to be a prehistoric survivor; a living pterosaur. It has been another of those serendipitous discoveries I make from time to time. About 2 weeks ago I went to Buxton, Derbyshire, U.K. to look around the town. Buxton is about half an hour by bus from Macclesfield where I live.

My friend and I went to Scriveners book shop in the town. It`s about 4 floors tall and whilst on the ground floor my eyes roamed joyfully along a shelf and I spotted *The Two Roads of Papua* by Evelyn Cheesman published by Jarrolds in 1935.

Cheesman (1881-1969) was a British entomologist and traveller. She was unable to train for a career as a veterinary surgeon due to restrictions on women`s education. Instead she studied entomology and was the first woman to be hired as curator at Regent`s Park Zoo, London. In 1924 she was invited to join a zoological expedition to the Marquesas and Galapagos Islands. She spent approximately twelve years on similar expeditions, travelling to New Guinea, the New Hebrides and other islands in the Pacific Ocean. In New Guinea she made a collecting expedition to the coastal area between Aitape and Jayapura (known as Hollandia at the time) and visited the Cyclop Mountains near Jayapura, collecting insects. Evelyn assisted at the Natural History Museum, London for many years as a volunteer. She was awarded an OBE for her contribution to entomology. At least one species of insect is named after her: the *Costomedes cheesmanae*

She also collected reptiles and amphibians and several New Guinea species were named in her honour: *Lipinia cheesmanae* (Parker, 1940 - a skink; *Platymantis cheesmanae* (Parker,1940) a direct breeding frog; *Litoria cheesmani* (Tyler,1964) a tree frog; *Barygenys cheesmanae* (Parker, 1936) – a microhylid frog;*Cophixalus cheesmanae* (Parker,1934).[1]

So I flicked through to see a few pages on mysterious lights and "ping" - I immediately thought "ropen".

The thing is a ropen has never, as far as I know, been seen by a westerner, but this light phenomenon of the grave-robbing ropen has been seen by both westerners and natives alike. This bioluminescence or whatever it is, is also associated with the flying snake of Namibia. [2]

On getting home I looked up the light in my copy of *Searching for Ropens* by Jonathan Whitcomb, 2nd edition:

Interviews by Blume [3] suggest that the bioluminescence may relate to secretions that seem to drip from the creatures as they fly, like "sparklers" falling to the ground. The secretions are said to burn human skin, sometimes seriously. In parts of Papua New Guinea the creature is worshipped. Some islanders on Umboi want nobody disturbing ropens, but on Manus Island people are simply curious about them. [4]

Cheesman saw the lights near a settlement of Mondo. Whitcomb does not mention Mondo in his book. This is to the northwest of the location where he and his fellow cryptozoologists were looking for the ropen, which was on the east coast of Papua New Guinea. But the "modern" ropen has been seen in mountainous areas of the country, such as Mount Tolo and Mount Barik, just as in Cheesman`s time:

"…Then it became evident what I had already supposed, that the flashes had been following a certain ridge of hills. Three ridges are visible one above the other in that direction, the highest one on the horizon. It was on the middle that this phenomenon appeared, and it seemed as if the flashes must have kept closely to the top of that one ridge. About a week later precisely the same thing occurred. There were the same sort of atmospheric conditions, except that this time the sky was rather cloudy…It may be dismissed at once that the flashes were due to any human agency…I could only imagine they were caused by some sort of gas escaping from crevices of rock". [5]

Interestingly Cheesman was told about similar lights in Queensland, again on a hill, and Whitcomb was told about a ropen near Perth, Australia in 1997. I e-mailed the address on the main website dedicated to searching for the ropen after seeing the book by Cheesman but so far I have had no reply.

(1)   Evelyn Cheesman Wikipedia http://en.wikipedia.org/wiki/Evelyn_Cheesman
(2)   R. Muirhead The Flying Snake of Namibia: An Investigation. CFZ Press. 1996
(3)   Jim Blume, eyewitness to ropen, 1996
(4)   J. Whitcomb. *Searching for Ropens* .2nd edition. 2007. p.104.
(5)   E. Cheesman. *The Two Roads of Papua*. 1935. p.226

SUNDAY, AUGUST 23, 2009

RICHARD MUIRHEAD: Snake stones in Yorkshire

Dear folks,

About two weeks ago I was looking in my late father`s garage for a collection of his doodles (as one does) when I came across a facsimile of part of an old British atlas, which I guess originally dated from the 17th century; the original map I mean not the reproduction. So I thought, well, this looks very interesting. I found a part of it covering the East and North Riding and read the following: (Original spelling adhered to)

"Places of memorable note are Whitby, where are found certain stones fashioned like Serpents, foulded and wrapped round in a wreath, euen the very pastimes of Nature, who when shee is wearied (as it were) with serious workes, sometimes forgeth and shapeth things by way of sport and recreation: so that by the credulous they are thought to have beene Serpents, which a coate or crust of stones had now couered all ouer, and by the praiers of S. Hilda turned to stones: And also there are certaine fields here adoining, where Geese flying ouer fall downe sodainlie to the ground, to the great admiration of all men:....At Skengrause (a little village) some seventie yeeres since , was caught a fish called a Sea-man, that for certaine daies together fedde on raw fishes, but espying his opportunitie escaped agiane into his waterie Element.....At Huntley Nabo, are stones found at the rootes of certaine rockes, of diures bignesse, so artificially shaped round by nature, in manner of a Globe, as if they had beene made by the Turners hand. In which (if you breake them) are found stony Serpents, enwrapped round like a wreath, but most of them headlesse".

I have asked the Bodleian Library if they can provide me with any information about this atlas but so far without success. Dr Darren Naish informed me that the Whitby and Huntley Nabo stones were ammonites and the latter were nodules, but William Corliss (who I rang) and I noted that the Huntley Nabo stones were globular. Indeed, I sent Corliss the information in this blog. He had never heard of such a thing. Strangely a day or two after I had found the facsimile map I found my dad`s doodles and he seemed to have specialised in ammonite-like drawings!

WEDNESDAY, AUGUST 12, 2009

MUIRHEAD'S MYSTERIES: A collection of cat curiosities

Muirhead`s Mysteries is back after a bit of an absence and today we take a brief look at cat curiosities. I touched on this in issue 16 of *Animals & Men* in an article titled ` A Collection of Cat Curiosities.` My apologies to anyone who has already published this information without my knowledge.

This first case really interests me.

All I have is the following bare note: *Naturalists Notebook* 1868 p.318. 'Flying cat. Shot by Alexander Gibson at Punch Mehab and exhibited at last meeting of Bombay Asiatic Society. Called by Bhells pauca billee. 18 inches long and as broad when extended.

Mr Gibson really believes it to be a cat and not a bat or flying fox as some contend.'

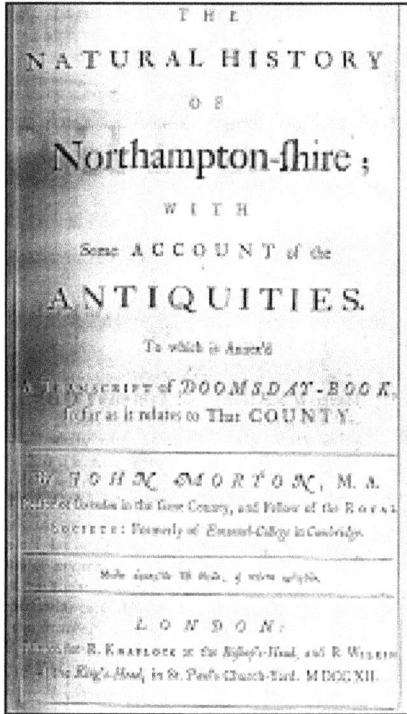

The title page reads:

THE

NATURAL HISTORY

OF

Northampton-ſhire ;

WITH

Some ACCOUNT of the

ANTIQUITIES.

To which is Annex'd

A TRANSCRIPT of DOOMSDAY-BOOK, ſo far as it relates to That COUNTY.

By JOHN MORTON, M. A.

Rector of Oxindon in the ſame County, and Fellow of the Royal Society: Formerly of Emanuel-College in Cambridge.

LONDON:

Printed for R. KNAPLOCK at the Biſhop's-Head, and R Wilkin at the King's-Head, in St. Paul's Church-Yard. MDCCXII.

The Sun of October 15th 1999 reported on Spike, Britain`s oldest moggie: 'A 29-year-old ginger and white tom-cat called Spike was yesterday crowned Britain`s oldest living moggie...She only discovered Spike was a record-breaker when she took him to a vet. She said: "I`d no idea his age was that unusual but the vet was staggered so I called the record people." Mo [his owner] added: "He must be lucky because he was bitten by a huge dog at 19. Vets didn`t think he`d live. "Britain`s oldest ever cat died in Devon in 1957, aged 34.'

I have several pages of information from *The Natural History of Northampton-shire* with some account of the Antiquities, etc, etc. (1712 p.443), which includes the following information on wild cats .

'Many Years ago we had wild cats in our Northamptonshire Woods; as appears by the Charter of King Richard I to the Abbot and Covent of Peterborough, giving them leave to hunt the Hare, the Fox, and the Wild Cat...And we now meet with them, tho` more rarely since the Woods have been thinned. These from their way of living, which is catching birds, on which chiefly they feed, are here called Birders. The wild Cat, that however of Whittlewood Forest, is generally larger Size, and has a Tail many Degrees bigger than the Tame. The wild Cats differ also in Colour from the common House-Cats...I mean in respect of the Colour, [of the Wild Cats] which for the main is a dusky Red or Yellow, and that in all of them; whereas in the Tame ones it is various and uncertain. The She Cats at Finshed, and the like Lone-Houses, do sometimes wander into the Neighbouring Woods and are gibb`d by the Wild ones there. `Tis a very difficult matter to the Wild Wood Cats, tho taken never so young into the House'.

Thus concludes Muirhead`s Mysteries for this evening.

MONDAY, SEPTEMBER 21, 2009

RICHARD MUIRHEAD: Snake with leg

Dear Jon,

Did you see this rather horrible thing?

Just because most mutants don't gain special powers doesn't make them any less interesting.

Case and point; this snake discovered the other day in southwest China. Looking at the picture, you should be able to figure out what makes this snake different from most; namely, the weird clawed limb sticking out of its side.

Dean Qiongxiu, the woman who found the snake, claims she discovered it stuck to the wall of her bedroom. Shocked and scared, Qiongxiu proceeded to beat the snake to death with her shoe, before preserving the beast in a jar of alcohol.

Obviously, this being a natural oddity from China with no independent verification, there's a good chance it's just a skilful taxidermy hoax. However, nothing in biology prohibits such a mutation.

JON: Personally I would suggest that this is actually a snake that ate a lizard that did protest too much, and shoved its leg through the stomach wall of the snake and out into the world outside.

SUNDAY, OCTOBER 11, 2009

A peculiar question from Richard Muirhead

JON: I have known him for 39 years, but never cease to wonder at the peculiar questions he has asked me with regularity during that time. Here is the latest (and as usual I haven't a clue what the answer is).

Hi,

Do you know if there are any records of giant tortoises in China during the Ming Dynasty, i.e. from 1368-1644AD?

Thanks,

Best wishes

Richard

MUIRHEAD'S MYSTERIES: Mylodons in New Zealand? Surely not!

Dear folks,

No, you are not hallucinating and no, it isn't April 1st. Whilst reading Gavin Menzies's book *1421: The Year China Discovered The World* (2002), which I mentioned in conjunction with a giant tortoise in Hong Kong, I came across comments about once-living mylodons in New Zealand!! The following is a copy of the notes I made about a year ago on this subject:

(This is the reason I made enquiries about New Zealand cryptozoologists in my blog the other day.)

'Several years ago…I made an interesting discovery, which may be relevant to the history of New Zealand's fauna. Menzies's theory is that huge Chinese vessels set forth from their home to reach much of the world, including South America and New Zealand via the east coast of Australia. Menzies says `The Chinese would have had to claw their way back against the current; [in the Tasman Sea] as they did so, at least two of the great treasure ships were lost. The wreck of an old wooden ship was found two centuries ago at Dusky Sound in Fjordland at the south-west tip of South Island. "It was said to be very old and of Chinese build and to have been "there before Cook" according to the local people.' [1]

This Chinese voyage would have taken place c.1421.

Furthermore: "Even more bizarre was a story, also reported to Collector of Customs in Sydney when the *Sydney Packet* returned home in 1831. One of the ship's gangs which had been stationed in Dusky Sound told of the discovery of an enormous animal of the kangaroo species.

The men had been boating in a cove in some quiet part of the inlet where the rocks shelved from the water's edge up to the bushline. Looking up they saw a strange animal perching at the edge of the bush and nibbling the foliage. It stood on its hind legs, the lower part of its body curving to a thick pointed tail, and when they took note of the height it reached against the trees, allowing a metre and a half for the tail, they estimated it stood nearly nine metres in height! The men were to windward of the animal and were able to watch it feeding for some time before it spotted them. They watched it pull down a heavy branch with comparative ease, turn it over and tilt it up to reach the leaves it wanted. When it finally saw them, the animal stood watching the men for a short time, then made one almighty leap from the edge of the bush towards the water's edge. There it landed on all fours but immediately stood erect before making another great leap into the water. The men were able to measure the first jump and found it covered eighteen metres. They watched the animal plough its way down the Sound at tremendous speed, its wake extending from one side of the Sound to the other." [2]

Menzies, commenting on this, says, "The animal described corresponds in size, posture and eating habits with the mylodons the Chinese could have taken aboard in Patagonia. Perhaps a pair escaped from the wreck, survived and bred in similar conditions to their home territories in Patagonia—the latitudes are the same. Sea-otters, which are not indigenous to New Zealand but, of course, were kept in the Chinese junks to herd fish, have been seen swimming in the fjords of South Island". [3]

Forty three years later, the *Otago Witness* reported on 'Notes of the Luna's trip.' on April 11th 1874, thus:

'There are two things, however, mentioned by Cook on which little reliance can be placed. One of these is that the sailors reported seeing a four-footed quadruped, with a bushy tail, in the bush, and from this statement of the sailors Cook concluded that there were land mammals at Dusky Bay. What this animal the sailors saw was, I know not. [4] Cook visited New Zealand in 1772. So if Menzies is to be believed, mylodons were supposed to have survived in New Zealand at least 351 years between 1421 and 1772, not to mention up to at least 1831'.

Part two will examine his claims of living mylodons further, otters on South Island, New Zealand, and other alleged animal travels from China's vessels to other parts of the world.

1. R.Gossett *New Zealand Mysteries* (1996) p.31
2. R. Gossett Ibid. pp148-149
3. G. Menzies 1421 *The Year China Discovered The World* (2002) p.173
4. *Otago Witness* Issue 1167, 11 April 1874 p.10

4 COMMENTS:

Dale Drinnon said...
Ground sloths were not leaping animals, and a leap of 18 meters is almost unheard-of even for a large kangaroo (that's sixty feet)

I would prefer some kind of a giant kangaroo to the ground sloth on that basis, but the whole story sounds a bit off to me. I will be glad to hear more. So far as I know there was some mention of some kind of unidentified "Wild pig or a cow" in Maori lore as well as what sounds like traditional descriptions of sheep and wolves.

Gary said...
Didn't know how to contact you other than by posting a comment - any idea what this is: http://4.bp.blogspot.com/_P09jpHp5b_U/StYkzHyLdBI/AAAAAAAACHU/79gc60imlfM/s320/seaton2.jpg

Nothing wrong with Europe '72 either - I saw The Dead on that tour.
cheers
Gary

Chris Clark said...
Some confusion between feet and metres surely? I can believe in a nine foot mylodon, but at nine metres it would be the biggest land mammal that ever lived.

Tilmeeth said...
I concur with Dale. Besides, weren't mylodons only found in the Americas?

FRIDAY, OCTOBER 16, 2009

MUIRHEAD'S MYSTERIES: Did a Chinese ship take giant sloths from Argentina to New Zealand? Part two.

Dear folks,

You may have read Part One of my blog on this subject; the farcical claim by Gavin Menzies that a Chinese vessel brought two living mylodons from the wilds of Patagonia to New Zealand. Firstly, how did the Chinese seamen capture them? How did they get them on board the ship? How can Menzies be sure the Chinese knew they were male and female? A website I found, called Gavin`s Fantasy Land (the author of the book who made the claim being Gavin Menzies) http://www.dightonrock.com/gavinsfantasyland.htm says:

'Why would the author even include such transparent nonsense,[that mylodon`s still live in Patagonia] resurrecting an extinct mylodon? (On p.120 of his book Menzies says- "…in recent years, well preserved pieces of this creature, apparently buthchered by the local people,have been found in a cave,leading to the speculation that it may still exist in the wilds of Patagonia.") We suggest Gavin doesn`t have any real evidence and he`s desperate. He`s setting up his "evidence" for a Chinese visit to New Zealand later, where a pair of these long extinct animals allegedly escape from the Chinese,and later on, he maintains, were the basis for a local legend. So therefore the Chinese must have picked up the mylodons first, as they passed by South America,so they could escape later in New Zealand.'

The author of this website then quotes from page 172 of Menzies book, which I mentioned in Part One of this blog, the "local people" who knew of the alleged wreck of a Chinese ship in Dusky Sound. The source cited in endnote 10 is Robyn Gossett, *New Zealand Mysteries*, Auckland,1996,p.31. Gossett explains it was not "local people" but the Maori who were keeper of the legend. Gossett devoted three pages to her detailed exploration of the legend, which included access to the log of Captain Robert Murry, who had been fourth officer on the ship. The wreck was not Chinese at all, but rather an English ship, the *Endeavour*, which went down in 1795.

So basically Menzies`s evidence that the Chinese took mylodons to New Zealand from Patagonia is based upon the flimsiest of evidence. Menzies also believes mylodons were taken to China from South America. A carving of an animal looking like a mylodon was dug up on the east coast of Australia near Gympie in 1966, thus supposedly proving Chinese visitation. On a website connected to Menzies books there are numerous references to animals allegedly taken by the Chinese from one part of the world to another: The remainder of this blog quotes from this website:

- Take chickens, As late as 1600 Mediterranean peoples did not have and did know of the galaxy of Asiatic chickens found in the Americas. Asiatic chickens cannot fly; someone took them to the Americas before Europeans got there.

- Dogs: Chihuahuas were really Chinese dogs that were imported by merchants. One of the supporting theories is that the Asians dwarfed animals and trees and may similarly have reduced the size of the Chihuahua. The flat furry tail, an important part of the Chihuahua is also common to other Asian breeds of toy dogs. The Basenji dog of Central Africa resembles the Australian and Thai dingos, and it was long thought Polynesians brought them from the Malay Archipelago to Madagascar. However recent DNA analysis suggests basenjis are closely related to dogs of Japan and China.

- Otters: domesticated otters, trained to fish, found in New Zealand, South Island, found in Ireland....Otters around the Isle of Skye—could they have been brought there by the Chinese fleets? More research needed.

- Camels to Peru

- Hippopotamus from Africa to China (Beijing Museum - `Western Han c.208BC`)

- Water buffalo to South America (Marajoara Island)

- Blue Magpie (*Cyanopica cyanus*) found only in southern Spain/Portugal and China and finally:

- Alligators - states that the species of alligator found in South Eastern United States is also found in but one other place on the planet; an isolated area along the Yangtze River in China. In every aspect other than size, they are identical though found 12,000 miles apart. Perhaps this is not an anomaly of nature.

These last two are complete nonsense. It has recently been discovered that the two populations of the azure-winged magpie (not the blue magpie, which is another species entirely) are actually distinct species, and the Chinese and American alligators have a number of morphological and genetic differences, and are clearly distinct species.

3 COMMENTS:

Retrieverman said...
Chihuahuas are Native American dogs with Spanish and other European dogs crossed in to make them tinier and apple-headed. They are very similar to the hairless dogs, which are all of Latin American origin. (Chinese crested dogs are NOT Chinese. They were actually marketed in the US first, as "African hairless dogs," which they probably also weren't. If there are any Chinese hairless dogs in history, they got there by the Portuguese and Spanish merchants who were responsible for

delivering them throughout the world. The actual breed we call a Chinese crested dog comes from a single breeding program in the US from the 1920's.)

And they are also similar to the hunting dogs that all sorts of Native Americans in the Northeast and Northwest kept. I have found several different Chihuahua type hunting dogs kept by Native American peoples in my research, among these are the Canoe Dogs, which look like Corgi/Chihuahua crosses (and were once said be evidence of Madog's discovery of America). These dogs just slightly larger than the Chihuahua, and were used to retrieve ducks by leaping out of canoes.

Another was the Tahltan bear dog, which I call the bear-hunting Canadian Chihuahua.

Tahltan: http://www.firstnations.de/img/05-3-0-dogs.jpg

These dogs were kept kept in the far North of British Columbia and the Yukon, and were most likely related the to the Canoe dogs, which I think are also related to the Chihuahua and Techichi-type dogs of Mexico.

Actually, the genetic evidence seems to point that all dogs have a Chinese origin, although this is certainly being debated rather hotly in academic circles. The last I saw about Basenji DNA is that they were related to Ugandan and Namibian street dogs than any from China:

http://news.bbc.co.uk/2/hi/science/nature/8182371.stm

This study is challenging the theory that dogs are from East Asia, which is very interesting.

However, I think it is unlikely that the small Native American dogs are actually Chinese. Native Americans were pretty good dog breeders. In Peru, they had big dogs for guarding their flocks of alpacas in the same way Europeans bred livestock guardian dogs to guard sheep and goats. In the Pacific Northwest, they had dogs that produced wool, which was then shorn to make yarn. The hairless dogs I mentioned earlier are very diverse animals, and judging from the art from the Pre-Columbian period, there were lots of diverse types of domestic dog.

Tilmeeth said...
Sorry, is it just me, but why on earth would the Chinese take animals from one part of the world and let them loose in another (going back to the Mylodon)?

And as much as I hate to our scorn upon a fellow Royal Navy and S&S veteran, Gavin Menzies doesn't even speak Chinese apparently... this website might also be of interest to those interested in this story http://tinyurl.com/yjlq37r

SATURDAY, OCTOBER 17, 2009

RICHARD MUIRHEAD: The Pygmy weasel

In the early 1990s I came across a reference to the 'pygmy weasel', an allegedly distinct mustelid; that is, distinct from the well known British weasel. This original reference was in *The Country Life* some time in 1975, which referred to pygmy weasels on Anglesey. My notes were passed to Jon and he incorporated them into his book *The Smaller Mystery Carnivores of The West Country* (1996), which is a must for any serious student of British cryptozoology. I quote from parts of Appendix Four of this book below.

Please forgive me for the frequent references to myself in this blog. I am not trying to "blow my own trumpet" it is just that am referred to quite a lot in Appendix Four of Jon`s book.

> `Richard Muirhead has spent some months investigating reports of dwarf weasels from various parts of the country, including the island of Anglesey and parts of Cumbria. The idea of such an animal is not particularly new. Writing in 1989 Sleeman noted: " The frequent existence of a second litter, coupled with the difference in size between the sexes are factors that give rise to stories about two types of weasels; ordinary and pygmy weasels existing side by side. In some rural areas such weasels are called `minivers.'`[1]

> Richard has discovered that these creatures are still widely believed to exist and in some areas are known as `Squeazels`. The details of the Anglesey animals are obscure, but it appears that they are lighter in colour than one would expect and have been reported as being white.'[2]

Jon then quotes from a letter I received on July 28th 1995. For the first time I reproduce the letter in its entirety; spelling, punctuation, etc is left as it appeared in the original.

> Dear Sir,

> I have been meaning to write to you for several weeks, re your letter in the Holyhead and Anglesey Mail asking for any information on the Pygmy Weasel.

> In my younger days I spent some time with an old mole catcher who worked the wooden barrel traps he occasionaly caught one of these Pygmy weasels in his traps. This was in the area of Church Stretton, Craven Arms and Ludlow area (Shropshire). The pygmy weasel is certainly not confined to Anglesy.

> There is a retired Game Keeper living near me who knew all about them when I mentioned it to him. They are to be found in Yorkshire Cumberland and Westmorland, he has also caught them in these areas from to time, and were quite a pest in the Pheasant pens (? word unclear) killing the young chicks and dragging them underground into the mole runs (?). The old mole catcher had a name for them (a Squeezel) due to pushing themselves down the mole holes.

I see one here on the farm from time to time, and recently saw one old one and three youngesters in a stone wall. There are two different types, the same size but one is quite a light colour and a much darker one is to be found. The darker being the rarest of the two.' [3]

On September 18th and 26th 1996 Colin Howes, the Environmental Records Officer of Doncaster Metropolitan Borough Council, provided me with further interesting information, which will be provided in Part Two, which will conclude the blogs on the pygmy weasel; along with information in a letter from Mr Paul Robinson on January 12th 1997.

1. P.Sleeman *Stoats and Weasels,Polecats and Martens*(1989) in J.Downes *The Smaller Mystery Carnivores of The West* Country (1996) p.111-112
2. J.Downes Ibid.p.112
3. Letter from P.M.A. Plews to R.Muirhead July 28th 1995.

> They call me hell, they call me Richard, that's not my name, That's not my name, that's not my name, that's not my name, they call me quiet boy, but I'm a riot, etc!!

(with apologies to The Ting Tings)

SUNDAY, OCTOBER 18, 2009

MUIRHEAD'S MYSTERIES: The Pygmy weasel part two

Dear folks,

In part one of this blog I introduced the pygmy weasel, which is either the female or juvenile stage of the common weasel *(Mustela nivalis)* or a different species altogether. In part two I introduce a few letters and articles ranging from 1935 to 2009 in order to provide more information.

On September 18th 1996 I received a letter on behalf of Colin Howes Environmental Records Officer of Doncaster Metropolitan Borough Council on the pygmy weasel. The letter read:

'Prof. Seaward of Bradford University has forwarded your enquiry to look into. During researches into the history of mammals in Yorkshire I have certainly come across allusions and possibly references to 'Pygmy'and 'Mouse' weasels but have simply regarded these as 'rustic' names to distinguish the common weasel *(Mustela nivalis)* from the stoat *(Mustela erminea.)* ...I seem to remember some of the gamekeeper correspondents of John Flintoff of Goathland, during the epic study of pelage and size in stoats and weasels earlier this century, referring to Pygmy weasels. Even today, members of 'shoots'in the Vale of Pickering region of North Yorkshire refer to very small weasels as `mouse` weasels." [1]

On September 26th 1996 I received a second letter from Howes about pygmy weasels
:

> Refering to Flintoff -'As an amateur naturalist Flintoff was well ahead of his time, undertaking a series of fascinating questionnaire surveys into the size ranges and ermine trends in stoats and weasels...Although Flintoff didn`t publish a book on the subject, his various papers are quoted in the Handbook of British Mammals and Dr King`s world monograph on stoats and weasels...The Game Conservancy at Fordingbridge, Hants* may be a useful contact for gamekeepers who would be able to provide the evidence you are looking for.' [2]

(*There is a note below this in my handwriting saying `They were not useful.`)

The R. J. Flintoff article that Howes sent with his second letter is `The Weights and Measurements of Stoats and Weasels `in *The North-Western Naturalist* for March 1935 vol 10 no.1

In Table 2 Flintoff presents the weight in ounces of three female weasels and fourteen male: 'The outstanding fact is that a weasel may weigh no more than 1 and a half ounces...Personally I hardly think the quality of size alone is sufficient as a basis for making a new species...Mr Adam Gordon, Duncombe Park, Helmsley, N. Yorkshire, an able and well known naturalist, writes: I must confess the weights of the weasels surprised me, because they were much less than I expected...Weasels are sometimes so small as to suggest there may be two kinds-the common weasel and the mouse weasel.' [3]

Coming right up to date with a web article dated September 18th 2009 from the *Darlington and Stockton Times* http://www.darlingtonandstocktontimes.co.uk `The mystery of the mouse-weasel and when a stoat becomes ermine` by Nicholas Rhea

'Rumours in country areas suggest there are two types of weasel. None of my sources confirm that two species exist in this country and yet the belief was once very strong, despite having no official backing... It is claimed that the further north one travels, the more likely it is that pure white stoats will be seen, although the famous colony in the ruins of Mount Grace Priory, near Osmotherley, have been known to turn pure white...So how has the belief arisen that two types of weasel live in England? In the none-too far distant past, it was claimed that there was a smaller variety known as the mouse-weasel. It was supposedly about 6ins (15cm) long, but in all other respects it matched the colouring and habits of the larger variety.

Some landowners claimed that the mouse weasel could pass through an average-sized wedding ring. Not surprisingly, experts suggested these small weasels were mere juveniles, but rural experts had considered that possibility and rejected it. So far as I am aware, this likelihood has never been satisfactorily answered, but it might be worth adding that a female weasel is smaller than the male, though only slightly. [4]

1. Letter from Colin Howes to Richard Muirhead September 18th 1996.
2. Letter from C. Howes to R. Muirhead September 26th 1996.
3. R. J. Flintoff The Weights and Measurements of Stoats and Weasels *The North-Western Naturalist* March 1935 vol.10(1) pp31,33.
4. N. Rhea The mystery of the mouse-weasel and when a stoat becomes ermine http://www.darlingtonandstocktontimes.co.uk

My next blog will cover some hitherto unpublished information from a German academic on Namibia's flying snake and news about my surprise appearance with the *Scissor Sisters* at a secret venue in Manchester. (Don't hold your breath, one of those comments is a joke!)

MONDAY, OCTOBER 19, 2009

MUIRHEAD'S MYSTERIES: The Flying Snake of Namibia

Dear folks,

The CFZ Yearbook 1996 published my article on Namibia's flying snake and with Jon's permission I hope to speak about it at the 2010 Weird Weekend. The *1996 Yearbook* covered the flying snake, or whatever it is, from mainly a purely cryptozoological point of view. I was aware of mythological aspects (I am aware of the fact that mythology and cryptozoology overlap) but at the time I was ignorant of the extensive research of Sigrid Schmidt of Hildesheim, from whose letters to me I quote in this blog.

When I sent Schmidt a copy of my 1996 Namibian flying snake article she/he was unfamiliar with the cryptozoological approach to the subject.

Schmidt's letter of September 16th 1995:

> 'Like the ghosts or UFOs in Europe, these snakes are seen by people who believe in them. And people who do not believe in them do not see them. A teacher once sarcastically told me: If at night people see the light of a motor-bike or a car where only one light is working people say: Oh, there is the snake again! And there are many people who delight to tell tales how they saw the snake or about other people who met the snake. Usually quite a number of different traits are attributed to these snakes, each narrator stresses different ones: its stench which alone kills people and attracts swarms of flies, its call which sounds like sheep or goats calling, the light, lamp, mirror, stone or white spot on its forehead, its face like a man's face, sometimes even with a beard, its horns or ears, its fondness for women. In dry Namibia the snake (which is usually called the Big Snake) lives in the mountains, but in the permanent rivers, particularly the Oranje River, its aquarian lives in the water, has a palace under the water and keeps there his human wives which he steals at the shore. These snakes

belong to a very ancient stratum of belief in Africa and in other continents as well. In southern Africa there are rock paintings of prehistoric times of huge snakes which probably were connected with rain or rain ceremonies

As to the flying snake in particular: Usually this snake has no wings but uses the end of its tail to push itself through the air to the next point. And as to the reporter of the 1942 accounts: the policeman Honeyborne was known as a very good narrator and experienced quite a number of extraordinary things.' [1]

Schmidt`s letter to me of October 15th 1995 makes a brief reference to crop circles near Hildesheim 'a few years ago.' and also: 'And as the main source for the 1942 report* was the policeman who was known as the great story-teller I just see no reason at all to accept it as reality in our sense.' [2]

The 1942 report of a flying snake in Namibia was the subject of my *1996 CFZ Yearbook* essay.

Finally, Schmidt`s letter of May 22nd 1997

"I believe that our good friend Michael Esterhuise experienced what he had learned from Nama sources,for his adventure corresponds exactly to Nama tradition. Legend students probably would call this "to act out a legend."'

Another remark: All the present-day studies of the subject focus on Sergeant Honeyborne`s report which, though written in the style of a police report, was written for a magazine and not for his office. As far as I know there is no official office report...A remark to the kind of flying of the snake...This is the peculiar way of flying by pushing itself off the ground. It corresponds exactly to the so called "shooting snakes" in German folk belief and particularly about the snake kings which often have a crown and are white. I have too little material to claim a dependence of this Nama belief from Central European sources. But it perfectly fits into the general scene: just in the area of the Northern Cape and Southern Namibia the influence of European folk belief on Nama belief ist enormously (sic).This influence originates from Afrikaans lore and is strongest among the mixed population. [3]

The thing that interests me here is why should Nama reality be any stranger than ours? Just because its Africa? Who should say that a folk belief that Jon has a gig with Crass in a chamber in the depths of a Devon river, or Lizzy meets Take That for a gig at the bottom of a lake in Lancashire be any more unbelievable than a group of witch doctors meeting a giant snake in "backward" Namibia?

1. Letter from S. Schmidt to R. Muirhead September 16th 1995.
2. Letter from S. Schmidt to R. Muirhead October 15th 1995
3. Letter from S. Schmidt to R. Muirhead May 22nd 1997

1 COMMENT:

Chris Clark said...
I can't find the 2006 Yearbook, but I remember this from Roy Mackal's book. He says that the boy heard a roaring noise and smelt something like burnt brass when he saw the snake, and passed out from shock and fright. The combination of roaring sound, strange smell and sensations of terror, followed by unconsciousness, suggest to me an epileptic attack. It would be interesting to know if the boy had a history of epilepsy (though people who have attack by no means always have another).

TUESDAY, OCTOBER 20, 2009

MUIRHEAD'S MYSTERIES: Jackals in Britain

Today I am going to take a chronological look at British jackals. All of this information is publically available but I thought I'd bring it altogether in one place.

Terry Hooper wrote to me on August 12th 1997: 'I am aware that the 1800-1900s saw many Jackal incidents. Animals brought back from India and even some bred (obviously) over here as jackal cubs were often sold on as fox cubs! Ditto Coyotes and wolves...These canines were described at times as wolves, which suggestes possible [sic] the Asian Jackal was involved.' [1]

Shuker comments: 'Interestingly, as noted by Alan Richardson of Wiltshire (*The Countryman,* summer 1975) an entry in the Churchwardens` Accounts for the village of Lythe,near Whitby, North Yorkshire, recorded in 1846 the sum of 8 shilling was paid for "One jackall [sic] head." As this was a high price back in those days, it suggests that whatever the creature was, it was unusual. By comparison, fox heads only commanded the sum of four shillings each at that time.' [2]

Karl Shuker also says '...When the supposed wolf responsible for several sheep attacks between Sevenoaks and Tonbridge in 1905 was shot by a gamekeeper on 1 March (*Times,* 2 March 1905), it proved to be a jackal.' [3] This incident is also recorded by Fort in Lo!:

The killing of sheep—the body on the railroad line— *Farm and Home*, March 16, that hardly had the wolf been killed, at Cumwinton, in the north of England, when farmers, in the south of England, especially in the districts between Tunbridge and Seven Oaks, Kent, began to tell of mysterious attacks upon their flocks. "Sometimes three or four sheep would be found dying in one flock, having in nearly every case been bitten in the shoulder and disembowelled. Many persons caught site of the animal, and one man had shot at it. The inhabitants were living in a state of terror, and so, on the first of March, a search party of 60 guns beat the woods, in an endeavour to put an end to the depredations...This resulted in its being found and dispatched by one of Mr R. K. Hodgson`s gamekeepers, the animal being pronounced, on examination,to be a jackal."

Queries, Answers & Correspondence

Correspondents will greatly oblige by writing on one side of the paper only.

Red Butterflies and Moths.—The caterpillars of many of these feed on dock and nettle that show or impart a red dye; one feeds on the ragwort. All, including the red admiral butterfly and tiger moths, are sometimes yellow, not red. It would be interesting to determine whether by changing the food plant for one with a yellow sap, the yellow varieties could be bred. The white markings on the fore-wing of a common tiger, *Arctia caja*, that I reared on an onion are yellow.—A. H. Silvrie. [By inbreeding for several generations and always feeding the caterpillars upon white deadnettle, numbers of the common tiger moths have been reared with yellow instead of red underwings.—Ed.]

Sweet Peas with Five Blooms.—In one of gardening papers last year, was an account of a sweet pea stem with "five blooms". This I thought was nothing very striking, but on my going to look at some named varieties growing here, I was not able to find one spike with five blooms, but plenty with four. This year I have looked them round again, and have found only three spikes with the five blooms. What has made it more amusing, every person I have asked if they have seen one like it, reply "yes, there are plenty of them," but generally have to "give up" with, "I ran find plenty with four." I wonder how many COUNTRY-SIDE readers have noticed the rarity of five bloomed stems?—J. Sharp, B.E.N.A.

Erratic Bees.—A swarm of bees made its appearance one morning this spring and settled on an elder bush a few yards from the hive from which it had emerged. The owner shook bees into a and left them for a while; suddenly, off they went and took lodgings in the roof of a lofty house two hundred yards distant, where they appeared to be lost to the owner for ever. After a time, however, they descended to the ground and entered a rat's hole. A box was placed over this, and a hole bored into the rat's dwelling, into which some smoke was puffed. This drove these erratic bees into the box, and they have since given no further trouble.—GAMEKEEPER.

A Country Club for Cyclists and Pedestrians.—Club runs are all very well, but cannot a club help its members to spend not only Saturday afternoons, but week-ends, and even week-day evenings, in the country? The motorist can spend ten guineas a year on a country club house. By substituting a marquee for a mansion, and meadows for lawns, the like luxury may be brought within the reach of all, and I am now submitting to the secretaries of field clubs, cycling and other clubs, a plan which enables members to have the advantage of a country club, with special reduced charge for sleeping accommodation on payment of a club subscription, which would vary with the number of members in each club, but would not exceed a few pence for each. This, for instance, because of its nearness to the river, to London and to beautiful country, is a neighbourhood of high prices. But a club of fifty members, by paying an annual subscription of a guinea (about fivepence a head), could secure for any member accommodation at this camp at the rate of 1s. 6d. per night (Saturday or Sunday nights 2s.) as often as he or she liked to come. Such is the power of co-operation. I shall be happy to send a booklet describing the camp to any one interested.—ARNOLD EILOART, The Camp, Ditton Hill, Surrey.

Dog-Fox Hybrid?—This curious animal, supposed to be a cross between a dog and fox, was killed some time ago wild, in a wood in Warwickshire. In colour and shape it resembled the fox very much, especially the hind quarters, as will be seen in the photo, the tail is thick and marked at the tip with white the same as a fox. When killed, it had very much the same scent; in size, not quite so large as a fox.—ARTHUR QUATERMAIN.

district. There is no use to "sprinkle drops on our caps and clothes," for the reason that the flies make for our faces—they would be welcome to sit six deep clothes! If eucalyptus be rubbed over the rush of flies is stemmed for moments, but during that time one's eyes are brimming over with tears, minutes later the flies come on in greater numbers, and even seem to smell of assistance in guiding them. eucalyptus is no use in the New Forest. one genuine remedy I obtained when bee-keeper's protective fluid, known as fuge, sometimes gives relief when rubbed the face.—J. BASSETT DIGBY, "Brackenhurst" Bournemouth.

Toad in a Nest.—In a farm garden Wigton, in Cumberland, I found a toad had taken possession of a vacated bird's and who certainly appeared to be quite home in his strange abode. By way of experiment he was very carefully removed secluded portion of the garden a dozen or so away. On going to the nest the next I found that Mr. Toad had returned and comfortably settled down as before.—REV. L. STEWART, Forest Gate, Essex.

Young Missel Thrush Feeds Young Blackbird.—In June I obtained a young cock missel thrush and brought it up; and a month I put a young blackbird into the cage. my great surprise the thrush hopped on perch, picked up a worm, broke it in pi and gave it to the blackbird.—J. H. Ein

"Pigs and Moon."—The of the moon considered pigs are killed mentioned in note on April 3rd is no exceedingly ancient and is religiously held by many country folk. bably the astrologers responsible for idea, and their ciples of also give credence to the The idea is the pigs should killed when light of the is increasing forcibly between the "first quarter and the "l moon" so the meat or bacon retains its substance and not shrink cooking; where if the pigs killed when the moon is on the wane the shrinks or runs to waste in cooking.—CLAU St. John, Cornel Wood, Dudley.

[A Photo.] [A. Quatremain.]

A Curious Animal.

This creature, supposed to be a cross between dog and fox, was killed in Warwickshire.

Flies in the New Forest.—In the discussion as to the best methods of driving away flies, I notice that one correspondent advises the use of eucalyptus. Now, among entomologists who spend summer holidays in the New Forest, the fly question is always to the fore. No one else can realise the myriads of flies which surround one in the Queen's Bower

Larks Singing when Perched.—I have often noticed that larks sing on the ground too the evening, especially if the weather is at a heavy or has been raining.—B. L. Hess Castle Street, Dover.

The story of the shooting of a jackal, in Kent, is told in the London newspapers. See the *Times*, March 2nd. There is no findable explanation, nor attempted explanation, of how the animal got there. Beyond the mere statement of the shooting, there is not another line upon this extraordinary appearance of an exotic animal in England, findable in any London newspaper. It was in provincial newspapers that I came upon more of this story. *Blyth News,* March 14. "The Indian jackal, which was killed recently, near Seven Oaks, Kent, after destroying sheep and game to the value of £100, is attracting in the shop windows of a Derby taxidermist " *Derby Mercury,* March 15 that the body of this jackal was upon exhibition in the studio of Mr A.S. Huthcinson, London Road, Derby". [4]

It would be interesting to find out if this taxidermist still exists in Derby.

The illustration below shows a possible dog-fox hybrid from an illustration in *The Country-side* September 28th 1907 p. 294. The accompanying text reads: 'This curious animal, supposed to be a cross between a dog and a fox, was killed some time ago wild, in a wood in Warwickshire. In colour and shape it resembled the fox very much, especially the hind quarters, as will be seen in the photo, the tail is thicker and marked at the tip with the same as a fox. - ARTHUR QUATREMAIN. [5]

Finally back to Mr Hooper on jackals: 'I`m trying to track down what happened to one that escaped from Sandbach, Cheshire in Winter 1961; also the one seen several times in the Delamere Forest area of Cheshire circa 1974. We know at least one was released near the Shropshire/Wales border in 1993/94 so we`re trying to find out if a farmer or car eventually got it.' [6]

1. Letter from T.Hooper to R.Muirhead August 12th 1997
2. K.Shuker *Extraordinary Animals Revisited* (2007) p.96
3. K.Shuker Ibid. p.96
4. C.Fort Lo! in *The Complete Books of Charles Fort* (1974) pp 666-667
5. A.Quartremain Dog-Fox Hybrid? *The Country-Side* September 28th 1907 p.294
6. Letter from T.Hooper to R.Muirhead July 18th 1998

> Cities come and cities go just like the old empires
> When all you do is change your clothes and call that versatile
> You got so many colours make a blind man so confused
> Then why can`t I keep up when you`re the only thing I lose?
>
> (The Scissor Sisters-`I Don`t Feel Like Dancin`)

7 COMMENTS:

Retrieverman said...
Very interesting.

The description of the dog-fox hybrid with a white tip on the tail made me think of the side-striped jackal. It is unique among the genus Canis in that it always has a white-tipped tail:

http://www.pixdatabase.com/data/t/h/e/theman/medium/2466-side-striped-jackal-canis-adustus-s2.jpg

Further, the animal's head shape does resemble that of a side-striped jacks, which has a shorter muzzle and a defined stop. It looks rather bulbous compared to the golden jackal.

The fact that this animal looks like it has a plume tail and breeches, like you'd find on a setter, even makes this argument stronger. Look at the tail and hindquarters: http://www.caninest.com/images/side-striped-jackal.jpg

The rest of these are most likely golden jackals, although this species is hard to distinguish from the coyote. Both can be sheep predators. Coyotes sometimes have a white tail tip, although this almost always indicate a tough of interbreeding with dogs.

Retrieverman said...
Here's one that really has the fox coloration: http://www.outdoorphoto.co.za/forum/photopost/data/517/Side_striped_Jackal.jpg

Retrieverman said...
And another: http://pro.corbis.com/images/AB006866.jpg?size=67&uid=BC5FCD9A-325F-499C-974E-6B412F4DBEBD

Retrieverman said...
It must have been a side-striped jackal cub, because it's smaller than a fox but very similar to one.

Retrieverman said...
Or it could have been a warrah pup:

http://www.messybeast.com/extinct/warrah.htm

Warrahs were semi-domesticated canids from the Falklands. We know very little about them. It would have made sense that someone would have brought one back as a pet.

They may be been in the genus Canis, because some genetic studies on warrah specimens suggest that they are more closely related to the coyote.

Considering the amount of time it took for animals to be transported in those days, a juvenile side-striped jackal or warrah most likely was born to captive parents.

Retrieverman said...
It can't be a warrah, though, because the warrah was extinct by 1876.

So it was probably a side-striped jackal cub.

ColinM1 said...
In my collection, I have a black & white postcard showing the Sevenoaks jackal after it was stuffed & mounted, with the story printed on the reverse of the card. It was photographed & published by Essenhaugh Corke & Co., but is presumably now out of copyright. Happy to send a scan for the website if this would be useful.

WEDNESDAY, OCTOBER 21, 2009

RICHARD MUIRHEAD: Birds behaving badly - beware! (Part one)

Hi folks,

Today I take a look at some examples of bizarre bird behaviour, including one or two snippets involving birds that do not fall into this category. As usual I work in chronological order. Tomorrow I present part two.

From *Country-Side Monthly* February 1911 p.57:

'Alleged Savage Gulls-A reader sends a cutting which describes how two fishermen, while cleaning their catch of fish "were attacked by a thousand ravenous seagulls and one of them received a number of wounds from beaks and talons before other fishermen beat off the birds with oars and boathooks." How much foundation of fact there may have been for this story would be hard to say; but some parts of it are manifestly incorrect. The gulls may have assembled in large numbers to eat the offal of the fish and may then have proceeded to steal the fishes also; but that they "attacked" the men is not conceivable. Each gull would be actuated solely by a desire to get food for itself; and collectively they would have no idea of getting the food by "attacking" its custodians first. The suggestion, again, of wounds from the "talons" of seagulls is absurd. Seagulls have small and tender webbed feet, and they would no more think of trying to strike a man with their "talons" than of tickling him with their tails-E.K.R.'

Unfortunately E.K.R. does not give a location for the above mentioned aggressive gull behaviour.

Jumping forwards to 1972, below is a very strange report from a very strange place, Hong Kong, where I spent 19 years, 1966-1985: Wanchai was the red light district when I lived there, on Hong Kong island. I don't know its fate since the colony returned to China in 1997. *The Star* was a tabloid type newspaper.

The Star, June 20th 1972. p.3

'Wanchai 'Ghosts' Play Tricks. The "ghosts" of three people feared killed in the collapse of an old Wanchai tenement on Friday night played some tricks overnight with rescuers. The search for the three had already been called off when a dove was suddenly spotted as it flew from the ruins. This sign of life electrified rescue workers, who re-started their frantic digging. Suddenly a car horn deep inside the ruins sounded - then stopped. "Sound the horn again if you're alive!" shouted a rescuer. The horn sounded again. The workers grabbed their picks and shovels with fresh zeal, ignoring the risk of a further collapse, and finally the rubble opened - to reveal an empty car, some bricks and wood resting on the steering wheel and horn. And in a nearby flat, the dove, which had earlier drunk greedily when given water, suddenly fluttered...and died.'

I don't know if the item below is significant or not, but I include it in case it is:

Fate October 1974 vol.27 no.10

'A Clue To The Bird. I believe I may be able to provide a clue as to the whereabouts of the photo of the giant bird nailed to the side of a barn. I saw the photo in question on television during the mid 1960s during a program on which Canadian author-columnist Pierre Berton was interviewing the late Ivan T.Sanderson who discussed the various oddities he had encountered during his career. I surmise that The Society for the Investigation of The Unexplained founded by Mr. Sanderson might have the photo - or it may be part of his estate. In any case it might be possible to check the videotape of this programme, for "The Pierre Berton Show" was syndicated by Screen Gems out of Toronto and it ran from about 1964 to 1973 on various television stations across Canada. No doubt Screen Gems keep files on all past programs - W. Ritchie Benedict, Calgary, Alta,Canada

The letters page in *B.B.C. Wildlife Magazine* in September 1987 featured several letters on the subject of 'Do birds mourn? (page number unknown)

'I was intrigued by the feature `Do birds mourn?` which has prompted me to write concerning an experience I had with mute swans at Loch Feachen, near Oban, in February 1986. My wife and I were heading north one morning when I spotted a diver, which we stopped to watch. I then noticed two mute swans which moved away from a group of about eight and headed for what looked like a dead swan on the foreshore. Walking up the road I found that it was a dead cygnet. I sat on the roadside and waited. The parents (I presume) reached the shore, and without leaving the water they stretched their necks across the shingle and began swaying to and fro. Though they made no sound,this was surely a lament. Lifting herself on to dry land the pen moved slowly towards her offspring, followed closely by the cob. As she reached the dead bird she ran her neck and face all over the body and carried out a last preening with her serrated bill. I usually steer clear of anthropomorphism, but I'm sure that anyone who had seen the two swans behaving like this would have been convinced that they were indeed mourning.

Sean Wood, Crowden, Cheshire.'

Now more aggressive gull behaviour, this time from *Country Life* September 7th 1995 (page number unknown)

'Life is imitating horror films in Sutherland, where avian aggressors are terrorising locals, a la Hitchcock`s film The Birds. Diving herring gulls have already lacerated several unprotected pates-one woman from the village of Embo required stitches- and locals in East Sutherland say they are being terrorised by the gulls. Skuas have long been recognised as aggressive birds, and those coming into close contact with them almost expect to be divebombed. But it is a mystery why gull species are behaving in this way." Edited by Tim Richardson.'

Finally, a handwritten note from my files:

'Large bird about the size of an eagle. Brownish with other markings. Ragged wings. At

Orcheston, Wiltshire.June 7th 1996.Orcheston is about 7 miles NNW of Salisbury. (I have no idea who gave me this information.)'

U2 `A Sort of Homecoming`

And you know it`s time to go
Through the sleet and driving snow
Across the fields of mourning
Lights in the distance…. (U2 `A Sort of Homecoming`)

WEDNESDAY, OCTOBER 21, 2009

RICHARD MUIRHEAD: Birds behaving badly - beware! (Part two)

Hi girls and boys; it`s me again.

All the entries below are from Wiliam Corliss`s book *Biological Anomalies: Birds* (1998) except the first entry, immediately below, which is from Doris Rybot`s *It Began Before Noah* (1972), pp. 96-97, a history of mankind`s association with animals, particularly in the form of zoos. Here is the rather heart-warming story:

'One of the oddest stories of strange bedfellows concerns an eagle housed in the Museum of Natural History in Paris. This was about 1784. The great bird was moping and refusing its food. The keepers decided that the only hope of encouraging its appetite and so preventing it from dying was to give it living prey. The victim chosen was an "English cock." He was put into the cage, and all stood round, hopefully waiting for the eagle to pounce. Instead, the eagle quite slowly approached the cock, looked him all over, and then-to the amazement of the spectators-spread a wing protectively over the smaller bird; and thus they walked together about the large cage.

The cock remained with the eagle, and from thenceforward it recovered its appetite for the usual dead meat, and was very soon completely restored to health. Simply, it had been moping for companionship. As our French author remarks: *Chose curieuse et combine instructive !'* [1]

Does anyone feel like following this up?

Concerning egg patterns, Corliss says:

'It is not uncommon for eggs to display… peculiar patterns and scribblings. Except as they might help in camouflaging the eggs, no profound importance can be ascribed to such random markings. On the other hand, once in a great while, an egg will be taken from the hen house with markings that are certainly not random and even seem to bear a "message"! J.Michell and R.J.M Rickard, in their Fortean classic *Living Wonders*, tell of several wondrous eggs. One egg making newspaper headlines in 1973 was clearly inscribed with a "6" on one end. In another instance, an egg bore the initials "WX". One more profound egg message announced "Jesus comes.' [2]

Now, Corliss on feathers:

'Feathers as weapons`. Some trogons, cuckoo-shrikes, and pigeons are armed with sharp pointed feathers on their backs and rumps. These feathers are partially erectile and are probably useful, porcupine-like, against predators. [3] If spine-like feathers are good defensively, they should have offensive capabilities, too. The cassowaries of Australia and New Guinea are, like all ratites, flightless. Their primary feathers have been turned into formidable weapons: spines some 28 centimetres long. Not only do these spines protect these large birds from abrasive vegetation [4], but they are also used in fighting. (Note, too, that the knife-like toenails of cassowaries can disembowel unwary humans.)' [5]

'Anting: Description. The vigorous, enthusiastic, and apparently pleasurable rubbing of the plumage with, or its exposure to, ants and other substances, such as mothballs and smoke. All of the animals, objects, and substances employed in anting are acrid or pungent. Anting behaviour often seems frenzied or blissful...Rooks/Burning matches. Even more intrepid was a tame Rook. This Rook, while with its former owner, Diana Ross, the novelist, took to opening boxes of matches of the non-"safety" kind. He quickly learned that by holding a match in his toes and pecking at its red head he could cause it to burst into flame. The moment this happens, Corbie, as the Rook is known, picks the match up in his beak, goes into a magnificent anting posture and rubs the lighted match up and down the inside of his arched wings.' (6,7,)

1. D.Rybot *It Began Before* Noah (1972) pp 96-97
2. J.Michell and R.J.M. Rickard *Living Wonders* (1983) p.167 in W.Corliss *Biological Anomalies: Birds* (1998)p.98
3. A.Thompson,A.Landsborough, *A New Dictionary of Birds* (1964) pp 153, 173,483 in W.Corliss Ibid p.45
4. F.B.Gill *Ornithology* (1990) pp 59,61,68 in W.Corliss Ibid p.45
5. See chapters BHX8-X2 in Corliss: *Humans III*
6. M.Burton " A Possible Explanation of the Phoenix Myth." *New Scientist* 1:10, June 27,1957 in W.Corliss op cit p.163
7. V.Markotic Current Anthroplogy 16:477,1975 in W.Corliss op cit. p.163

That`s all, folks. Next blog will look at odd-coloured foxes in Britain.

> They tell us that we lost our tails,
> Evolving up from little snails,
> I say it`s all just wind and sails,
> Are we not men? We are Devo! *(Devo-`Jocko Homo`)*

FRIDAY, OCTOBER 23, 2009

MUIRHEAD'S MYSTERIES: Fleet foxes? Er, no, dummy, Fortean foxes!

Dear folks,

Here`s another selection of notes from Muirhead Archives in Muirhead Mansions. Actually 'Fortean foxes' is a bit of a misnomer because I am only looking at odd colouration in foxes today, not the whole range of Fortean possibilities in foxes. I hope you enjoy the following and if you have any more observations please can you contact Jon or myself? Thanks.

I found a note in a book called *The Sedgefield Country in the Seventies and Eighties* (and that meant the 1870s and 1880s) about a male *blue* - yes *blue* - fox in Mainforth-whin, in what I thought was the West Country. But I have just done a Google search on the names Sedgefield and Mainforth-whin and I could only find a Sedgefield in Co. Durham, once Tony Blair`s constituency. This is the only instance of a blue fox in Britain that I am aware of. The next time I go to the Bodleian Library in Oxford or the British Library I will try to look at this book, which I found in a charity shop in Taunton earlier this year.

Some time in 1995 Jan Williams (through whom I 'rediscovered' Jon after our time in Hong Kong) wrote to me about albino foxes and other things:

> "A pure white fox was killed in 1887 by Taunton Vale Hounds in West Somerset." This is in *Man and Beast* by Ron Freethy (?) Blandford 1983. [1]

In *Country Sportsman* 1949 (page number unknown) there is a story: `Albino Foxes In Northumberland A Strain That Persists Around Rothbury`

> `In the year 1937 a white fox was killed in the grounds of Brinkburn Priory almost within sight of the River Coquet by the Percy Hounds. The mask is now believed to be at Alnwick Castle*, the property of the Duke of Northumberland. In the spring of the same year, Richardson, a keeper on the Cragside Estate further west along the Coquet Valley, dug out a white vixen which had two cubs of normal colour. On Tuesday,15th 1938, the Morpeth Foxhounds, whilst hunting in the east part of their country, roused and quickly killed a pure-white fox. The mask, beautifully mounted, is now at Meldon Hall. The tips of the ears and brush are black, the eyes yellowish and lighter than the eyes of a normal fox, and there is none of the pink colouration one associates with the true albino....Later in the season another white fox was reported as being seen by the rabbit catcher at Paxton Dene, but no further trace of it was found. This outcrop of albinism naturally caused a good deal of local interest and there were many wild and fantastic theories as to the cause of this phenomenon. One of the most popular was that these white foxes had all been fathered by a silver dog fox which, by a strange coincidence, had escaped early in the spring of 1937 from a silver-fox farm in the neighbourhood of Capheaton,not very far, as a fox will travel, from the Coquet banks....In 1947 the keeper at Linden Hall, which lies about a mile north of Paxton Dene, reported having seen a white fox in the Dene. This man`s

evidence can be taken as reliable.....The seed of this white breed, I feel sure, originated somewhere amongst the rocky, rhododendron-grown hills above Rothbury and, from time to time ,it keeps cropping up as it is handed down from father to son.' [2]

* Coincidentally I will be at the Franciscan friary in Alnmouth near Alnwick in mid-November so I will see if I can see the mask at the Castle.

Jumping forwards 38 years to *The Mail on Sunday* August 30th 1987, `Lair of the little white foxes` by Dr Brendan Quayle, which told of with two white cubs, one of which was shot the other "booted to death" [3]. 'According to David Bellamy* the birth of an albino or white-skinned specimen of a wild or even domesticated creature, unless specially bred, is very rare.' [4] One fox was shot. The story continued concerning the killing of the other fox: '...It was two teenage sons of a local farmers who killed the other white cub and one of its red brothers from the same litter. "Why did you do it?" I asked them bitterly this week. "Foxes are vermin and we were worried they were going for our geese," they told me.' [5] Quayle speculated their white pelts would end up at a taxidermist.

* I wrote to David Bellamy four or five months ago about this story but received no reply. The above incident was in the Border country between England and Scotland. Perhaps the same location as in the Country Sportsman article?

According to Roger Burrows in *A Complete Study of The Red Fox* (1988) :

'There are records of white, presumably albino, foxes from Dartmoor, and at least five records of them from Whaddon Chase."... "Russian authors mention blue and silver foxes as being present in the northern of the red foxes` range, so not all the blue foxes in northern Europe need necessarily be descended from the feral North American form.' [6]

The *Wild About Forum* on the Net reported an eyewitness sighting of a white fox near the Wirral. This was on May 2nd 2006. 'Lemming' thought it was a dog but later:

'..This morning I saw it streaking across the field in front of my house, grabbed my binoculars and saw it was a white fox. White from head to hind legs then it turned a peachy colour. How lovely!!! Just wanted to share my sitings [sic] with you all.' [7]

On the same Forum in September 2007 there was some interesting communication about black and other coloured foxes. On September 22nd John (in Coventry) said: 'I have just had a shock. I was looking out of my bedroom window and a scrawny looking Black Fox walked past my Bungalow...I have never seen a Black Fox before and must admit that I didn`t know they could be that colour. It wasn`t a full on black but dark enough to be able to pass for black.' [8]

C C replied giving instances of a black one in Maesycmmer near Caerphilly and one on September 21st 2007 'on the fields by Carmarthen Bay'with 'a black stripe down back, with a stripe running down shoulder blades.' [9] C C saw three silver foxes in the past four years in Carmarthen.

Finally, on September 18th 2008 a newspaper website reported the sighting of a black fox on the outskirts of Chorley, Lancashire: Mr Hehir, from Preston, Lancashire, was walking in a cemetery with a friend when he spotted the animal among the gravestones.....Country villagers traditionally told stories of how the fox was as 'black as night, so that it could live in a man's shadow and never be seen.' [10]

REFERENCES

1. Letter from Jan Williams to R.Muirhead c.1995.
2. *Country Sportsman* 1949.
3. *The Mail on Sunday* August 30th 1987 `Lair of the little white foxes`.p.13
4. *The Mail on Sunday* Ibid p.13
5. *The Mail on Sunday* Ibid. p.15
6. R.Burrows. *A Complete Study of The Fox.* (1988) pp71,73
7. http://www.wildaboutbritain.co.uk/ May 2nd 2006
8. www.wildaboutbritain.co.uk September 22nd 2007
9. Ibid September 22nd 2007
10. http://www.telegraph.co.uk/ September 18th 2008.

That's all for today, friends. Tomorrow: `Talking Turtles.`

5 COMMENTS:

Retrieverman said...
Fur-farmed domestic red foxes often come in these colors.

This captive-bred red fox seems to meet the description of some of these animals really well: http://www.youtube.com/watch?v=8SgJYbRFrd0

There are the Belyaev foxes, too, although these are from the Soviet union: http://www.youtube.com/watch?v=enrLSfxTqZ0

In the native North American population of red foxes, we have lots of black and silver foxes. We also have an intermediate color, called a "cross fox." http://www.ejphoto.com/images_UT/UT_CrossFox07.jpg

It is believed that black and silver foxes existed in the original European red fox population, but these were killed off for their fur early on. In most of the Eastern US, the red fox that lives there is not the native red fox, but a derivative of exactly the same subspecies that exists in England. In fact, the two probably should be considered the same subspecies: http://bss.sfsu.edu/holzman/courses/Spring%2005%20projects/RedFoxCasey/RedFox02.jpg

The subspecies fulva is the one derived from the English red fox, which was imported so that English colonists could go riding to hounds. That particular area was originally home to only a few red foxes and lots of the species called gray fox, which is not a Vulpine fox but very primitive canid in the genus Urocyon. It is not a good thing for hounds to chase because they retain the primitive canid talent for climbing trees. In parts of Latin America, it was called the Gato Cervan "deer-like cat," because they knew it couldn't be dog. Dogs can't climb trees.

Also, one cannot ignore the fact that Arctic foxes were fur-farmed (and still are). The most common phase of the Arctic fox in the fur industry is the blue fox.

In the summer, blues look like this: http://www.ejphoto.com/images_AK/AK_ArcticFox17.jpg

http://www.corbisimages.com/images/DG004430.jpg?size=67&uid=9C0A484F-DBB3-4191-9F8F-67D6E0088909

Unlike silvers, blue foxes are actually a different species from the red. The diagnostic for a red fox is that the animal with have very long legs and a white tail tip. (Although that cross fox photo shows one with a very limited white tip!)

Although Arctic foxes are more closely related to the swift and kit foxes of North America (and Arctic foxes have since been moved from the Genus Alopex to the Genus Vulpes to reflect this), red (silver) and Arctic (blue) foxes have hybridized in captivity:

http://www3.interscience.wiley.com/journal/119853351/abstract?CRETRY=1&SRETRY=0

I think this is a good post, but one of the experts in it got confused about the blue foxes. The blues are a separate species-- a color phase of the Arctic fox.

I've generally read that most silver and black foxes (which are red foxes) in existence today come from Atlantic Canada, so if there is a wild Russian population, that is certainly fascinating.

Retrieverman said...
The one with a black stripe down the back sounds like a cross fox.

They are called cross foxes because the black stripe that runs down their backs and another that runs up from their black legs across their shoulders, forming a cross on their backs:
http://www.westford.com/fingerhut/Alaska/Cross-Fox.jpg

Retrieverman said...
In captivity there are several different colors of the red fox now being bred-- too many colors to count. Some of the blue Arctic foxes have some white markings: https://91middleschoolscience.wikispaces.com/file/view/Arctic_Fox.jpg (This is in winter coat).

Some blue cubs with those markings (summer coat): http://bioweb.uwlax.edu/bio203/ s2008/olson_alex/reproduction.htm

This is what they normally look like: http://www.treknature.com/gallery/ photo181526.htm

This litter of Arctics has blue and normal (white) phase cubs: http://cache1.asset-cache.net/xc/dv425013.jpg? v=1&c=IWSAsset&k=2&d=EDF6F2F4F969CEBDB74E030E642B174CD2CA2C7EDC 48C69434C488367188741FE30A760B0D811297

The white ones look a lot like swift foxes in the summer: (Arctic) http:// retrieverman.files.wordpress.com/2009/04/white-arctic-fox-in-summer.jpg

(Swift):

http://www.ics.uci.edu/~eppstein/pix/josh3/SwiftFox-m.jpg

The swift and Arctic fox are very closely related. I don't know if the Arctic is a swift fox that lives in the tundra and frozen ice or if the swift is an Arctic fox that lives on the prairies. They are also related to the kit fox of the Southwest, which is adapted to living in the desert. http://www.canids.org/gallery/Popup19385.jpg

They have been known to interbreed with Swifts, and at one time were considered the same species.

Retrieverman said...
Now, there is also a platinum coloration of the red fox, which appears blue: http:// www.billsbearrugs.com/Inventory2008/Platinum_Fox%281%29.jpg

I have never heard of this color in wild red foxes, so it is most likely a fur-farmed variety. Maybe there are or were real platinum foxes in the wild in Russia.

The only colors I've heard of in the wild are the normal red, the cross, and the black/ silver.

The blue and white Arctic varieties do exist in the wild, though.

Retrieverman said...

When Arctic foxes transition from one season's coat to the other, they look bizarre:

1. http://www.papiliophotos.com/SearchImages/P-MAM095-54.jpg
2. http://www.gvzoo.com/files/u2/Arctic_Fox_093_cropped.jpg
3. http://farm4.static.flickr.com/3087/2694994098_9e4d16c3eb.jpg
4. http://homepage.mac.com/wildlifeweb/mammal/arctic_fox/arctic_fox01.jpg
5. Some blues look chocolatey brown: http://www.naturetrek.co.uk/pics/t346-large2.jpg
6. http://blogs.nationalgeographic.com/blogs/news/chiefeditor/Arctic-Fox-with-goose-egg-picture.jpg
7. http://sierrabirdbum.com/Mammals/Arctic-Fox.jpg

SATURDAY, OCTOBER 24, 2009

MUIRHEAD'S MYSTERIES: Turtle tales

Hi folks,

Today I'm looking at mysteries involving turtles, that is, cryptids and odd appearances, delving into the realm of folklore. Also, the information presented here cannot be presented in a strict chronological order because the starting point or origin of some of these cryptids in their impact on mankind cannot be accurately discerned.

Firstly, Dr Roy Mackal in his book *A Living Dinosaur?* (1987) records an animal the locals call `Ndendeki` - a giant turtle:

> 'Informants may have felt it was quite enough simply to say "giant turtle" about an animal whose shell was some 4 to 5 metres (12 to 15 feet) in diameter. Such a giant turtle hardly seems believeable, yet the fossil record includes both marine (*Archelon ischyros*) and land turtles (*Colossochelys atlas*), their remains indicating animals 6 metres (20 feet) in length. However, not even the giant tortoises of the Seychelles Island (1.2metres, 4 feet) approach the reported dimensions of Ndendeki....Based on reported descriptions, we require a very large, freshwater turtle, both carnivorous (or omnivorous) and amphibious. Marcellin Agagna was able to identify the Ndendeki as one and the same with *Trionyx triunguis*.'[1]

Michael Newton, in his *Encyclopaedia of Cryptozoology,* records a creature called the Carvana. '

> This insidious predator of swamplands and lagoons in eastern Texas was described by a Mexican migrant named Aluna, who allegedly lost livestock to the creature in the mid nineteenth century. According to Aluna`s story, no Caravana was ever seen alive, though its

remains were sometimes found during droughts, when its marshy habitat evaporated. On those occasions, local settlers and aboriginal tribesmen allegedly discovered skeletons resembling those of huge turtles, with shells 10-12 feet long and 6 feet wide, while the beasts head and tail resembled an alligator's. In life, the Cavanar lay submerged in mud, waiting to pounce on prey that included livestock and human beings. Skeptics suggest that tales of monsters "never seen alive" refer to early discovery of fossil dinosaurs or Ice Age megafauna.' [2]

I now come onto folklore. In the late 1990s I bought a book called *The Legend of A Giant Tortoise* number 331 in the Asian Folklore and Social Life Monographs published by the Chinese Association For Folklore in Taiwan (1997). It was sent to Dr. Richard Muirhead (I have many faults, but I don't think delusions of grandeur are one of them except when I'm ill!) at 9, The Grift near Salisbury when it was actually 9, The Croft. Anyway, most of this book is in Chinese, but there is enough in English to make it worthwhile to own. I cannot speak or read Chinese. On page 77 there is a long list of folklore motifs connected with `Tortoise, see also Turtle` and Tortoise's...etc`. Some are quite amusing if you have my kind of sense of humour:

`Tortoise, see also Turtle....breaks elephant's back...catches ogre helping him....has no liver or teeth...speaks and loses his hold on the stick....fight between ape and tortoise.....man transformed to tortoise...tortoise's foolish association with peacock [I wonder what that's all about?-RM] ....why tortoise's neck is outstretched to sky; is looking for his wife,the star.` [3]

Page 55 of the above-mentioned monograph gives an extract from Bernard E. Read's translation of a Chinese *Materia Medica*:

'Lu Mao Kuei Green Haired Tortoises (a) Comments. Li Shih-chen. The green haired tortoise comes from Nei hsiang and T'ang hsien. In Li's day they came from Ch'i chou (Hupeh) and were used as presents. Taken from the streams they were reared in large jars and fed with fish and shrimps. In winter the water was removed and after a long time they grew hair four to five inches long with some golden threads. The genuine article has a carapace with three ridges, the plastron is as white as ivory, and it is as large as the Wu Chu Ch'ien...Han Dynasty money. Other turtles if kept a long time grow hairy but they are larger, the hair has no golden threads, and the plastron which is yellowish black shows a difference. The Nan Ch'l Shu ....states that in the time of Yung Ming (A.D. 483) there were presented to the court green haired sacred turtles , which refer to this turtle. The Lu Yi Chi .....states that in Hsuan Tsung's time (A.D. 713) a Taoist presented to the court a small turtle one inch in diameter , golden colour, very lovely in appearance, it being said that placed in a bowl it would give protection from poisonous snakes. This is one of the marvels among turtles... Qualities. Sweet, bland, acid, non-poisonous. Uses. It increases the "Jen-mo" circulation (aortic?) It stimulates sexual desire, and menstruation. A seminal tonic. It increases erection. Ch'en Chia mo. Tied on the forehead it will prevent an attack of malaria; and placed on the book case it will keep away bookworms.' [4] I HONESTLY didn't make up the second part of that sentence! So, Jon, Lizzy, remember not to place a hairy tortoise shell on your bookcases tonight!

綠毛龜

Figure 2

Cm.

4 Kc.

(a) Giles and Williams both state this is the green...terrapin from Szechuan on which a species of conferva grows. Two algae are said to grow on the *Geoclemmys reevesii*, Gray viz, the *Cladophora glomerata* and *Basicladia crassa*. Abs from Chinese Science. Soc. Proc. Nanning 1935. Aug.

The image is from Read`s *Materia Medica* showing the green-haired tortoise.

Around about the time I bought this monograph there was a story and photo in the *New Scientist* of a Chinese tortoise with hair growing from its shell.

Finally I look at Moka Moka, a giant turtle from Queensland. I couldn`t find it mentioned on the Web nor Newton's above-mentioned Encyclopaedia:

"Gigantic Sea Turtle." Examples of this very large species of marine turtle have been recorded in areas of the South Pacific, and some fine skeletal remains have been recovered in Australia. Relatively current reports say that in Queensland there is a massive turtle known as Moka Moka. One report by a young woman named Lovell, a resident of that area, was recorded in 1890. She said that the sea turtle had either teeth or serrated jaw bones. She went on to state that "what I saw of it was about 27 or 28 feet, but I think it must be 30 feet in all. Whilst its head was out of the water it kept its mouth open,and, as I could not see any nostrils, I fancy it breaths through its mouth. The jaws are about 18 inches in length ;the head and neck greenish white, with large white spots on the neck, and a band of white round a very black eye." (5)

1.   R. Mackal. *A Living Dinosaur?* (1987) p267, 269
2.   M. Newton *Encyclopaedia of Cryptozoology A Global Guide* (2005) p 89
3   Y. Hung et al *The Legend of a Giant Tortoise.* (1997) p.77
4   B. Read *Chinese Materia Medica* (1977) in .Y.Hung et al Ibid. pp54-55
5.   J.Sweeney *A Pictorial History of Sea Monsters and Other Dangerous Marine Life* (1972) p.137

Tomorrow- The Sydenham, London, panther.

And all the world is biscuit-shaped,
Its just for me to feed my face,
And I can see, hear, smell, touch, taste,
And I`ve got one, two, three, four, five,
Senses working overtime
Trying to take this all in
I`ve got one, two, three, four, five,

Senses working overtime   (*XTC* `Senses Working Overtime)

1 COMMENT:

Retrieverman said...
Turtles are always interesting. I've never seen any giants, but North American freshwater turtles are very hard to identify.

Guide books (like the kind used for birds) don't really help when the animal is covered in algae.

SUNDAY, OCTOBER 25, 2009

MUIRHEAD'S MYSTERIES: Mystery cats of south-east London

Hi folks,

A few days ago my brother Bill, who lives in Sydenham with his wife Jane, passed on to me a newspaper cutting from *London Life* dated September 28th 2009 (see below) on the `Sydenham panther`, which reminded me that I had come across a mention of this creature a few years ago. So I have just done a fairly quick bit of research on the history of this mystery cat. It transpires that since at least 2005 there have been a series of sightings in various locations (perhaps of more than one animal?) around towns in southeast London with Sydenham and Crystal Palace (where one is known as the Palace Puma), where there is an open park space as the 'epicentre' as it were. There seems to be a gap in reports between March 2005 and September 2009, at least as far as the Internet is concerned.

On March 23rd 2005 *The Guardian's* web site reported: Fear stalks the streets of Sydenham after resident is attacked by a black cat the size of a labrador.` By Patrick Barkham.

theguardian

Fear stalks the streets of Sydenham after resident is attacked by a black cat the size of a labrador

Man was calling to pet when 'panther' struck

Patrick Barkham

It has probably slunk off to a neighbouring suburb to become the Penge Panther, the Catford Cheetah or the Beast of Beckenham by now.
But residents of the blossom-filled streets of Sydenham were still shaking last night as a father of three told how he had been mauled by a black cat the size of a labrador.

Police armed with Taser stun guns sealed off roads in south-east London, school ...locked and teachers warned pupils to keep away from wooded ar... ...lder escaped with a cuff around th... face from the big c...
...s calling in ... .15am v...

'It has probably slunk off to a neighbouring suburb to become the Penge Panther, the Catford Cheetah or the Beast of Beckenham by now. But residents of the blossom filled streets of Sydenham were still shaking last as a father of three told how he had been mauled by a black cat the size of a labrador. Police armed with Taser stun guns sealed off roads in south-east London, school gates were locked and teachers warned pupils to keep away from wooded areas after Tony Holder escaped with a cuff around the face from the big cat...As Mr Holder was being treated for scratches by ambulance staff, he saw the beast saunter past again...The animal gave them the slip, but as tabloid reporters scoured the streets in safari gear brandishing butterfly nets, the Guardian picked up the scent of something big across the railway line by Catling Close. Billy Rich, 44, was looking out of his window at 5.30am when he saw a black creature leap across the road and bound south towards Mayow Park...Scotland Yard confirmed the beast of Sydenham was the second serious sighting of a large black cat in south-east London in the past three years. Officers responded to reports of a large black cat in Oxleas Wood in Shooters Hill, south London, in 2002 but failed to trace the animal' [1]

We now jump forward about 4 and a half years.

StreathamGuardian.co.uk reported on September 1st 2009

'It seemed like any other day for Helen Barrett when she took a woodland walk with her family earlier this month [presumably August 2009]. But that changed when she claimed the 5ft Palace Puma stalked out of the undergrowth and fearlessly approached her. Understandably the 41-year-old turned and fled, later describing her encounter in Crystal Palace as "alarming".' [2]

So we now have a black panther and a puma. This site also mentions sightings in Eltham, Sidcup, Dartford, Gravesend, Bexley and Thamesmead and mentions the CFZ's Neil Arnold's research.

The *London Life* report of September 28th 2009 was headlined 'Sydenham Panther on the prowl' by Mark Blunden:

'A panther is feared to be prowling south-east London for new victims today after a pet cat was found mauled to death. The remains were discovered in the neighbourhood where the so-called "Sydenam Panther" is believed to have struck four years ago. The dead cat was discovered by Sara Hill, 32, as she walked her dogs with her son, Archie, eight months. She said the feline skeleton had been entirely stripped of flesh - yet whatever killed it had left the head untouched.' [3]

On September 30th 2009 a BarnacleBelle on the Sydenham Town Forum website [4] mentioned the lion of Lewisham and the Honor Oak Hyena, both of which I'd like to follow up but I haven't got the time.

That was essentially the end of sensible debate on this forum until a contribution by coral on

October 16th 2009: 'A month ago whilst walking home along Datchet Road, Catford between 22.00 and 22.30hrs I saw what appeared to be a big cat. It was strolling between cars and front pathways of houses. I had no idea of previously reported sightings of black panthers and such like until my daughter later informed me of such. It was not black but more a brownish colour. The height was about 2-3ft high, thin body frame and its fur was much shorter than a pet cat. The head was definitely that of the cat family. Make no mistake about it all you doubting Thomas's out there, a big Cat is definitely at large in borders of Catford/Sydenham.' [5]

This sounds a bit like the Palace Puma.

1. Guardian website. http://www.guardian.co.uk/2005/mar/23/patrickbarkham
2. StreathamGuardian.co.uk website. http://www,streathamguardian.co.uk/ news/4571308./
3. M. Blunden Sydenham Panther on the prowl. London Life September 28th 2009 p.23
4. Sydenham Town Forum website September 30th 2009 http://sydenham.org.uk/forum/ viewtopic.php?=30490
5. Ibid. October 16th 2009

That's all for today. Tomorrow I'll be looking at the case study of an entombed toad.

Fishing children fish in the Imperial Waters
Nagasaki nightmare, Nagasaki nightmare
Sons and lovers, lovers and daughters
Nagasaki nightmare, Nagasaki nightmare
Cherry blossom hanging on the cherry blossom tree
Flash, blinding flash, then there's nothing to see..
Crass, Nagasaki Nightmare`

2 COMMENTS:

Neil A said...
Hi Richard,
There is no 'Palace puma'; this was inaccuracy down to the press. There was no attack by a 'big cat' on a Mr Holder but a suspected hoax.

I devoted a whole chapter to the Sydenham turn-out and other London mystery cats in my new book 'MYSTERY ANIMALS OF THE BRITISH ISLES: KENT' which will continue when I submit my London book.

There is no 'Lewisham lion' either. All 'lion' or 'lioness' reports from the capital are unfounded...i.e. 'Winchmore Hill lioness', and also the 'Edgware tiger'...

There is certainly at least one black leopard which prowls from Norwood, to the outskirts of Kent at Bexley, Bromley, but sightings have also taken place in nearby

Surrey - Croydon for example.

For further info go to my: www.beastsoflondon.blogspot.com for full details.

Tilmeeth said...
From the Streatham Guardian article: "...and causing much alarm amongst local ramblers."

What a fantastically British thing to say! I liked the accompanying video to, not quite what any of us were expecting perhaps...

The Guardian link should be http://www.guardian.co.uk/uk/2005/mar/23/patrickbarkham

And the forum post link should be http://sydenham.org.uk/forum/viewtopic.php?t=3915 regardless, very interesting and informative post. I look forward to more of Muirhead's Mysteries...

TUESDAY, OCTOBER 27, 2009

MUIRHEAD'S MYSTERIES: Case study of an entombed frog

Dear friends,

There is nothing new about entombed toads; they are well recorded in the annals of Forteana. However, this is the first time I have come across an entombed frog. So in this case study I present information that might be new to you. It was kindly provided to me a long time ago by the father of my friend Rob Wilkes. If anyone is interested in following it up further the *Dean Forest Mercury* may be in the Newspaper Library in Colindale, N. London.

The correspondence and articles in the *Dean Forest Mercury* is in the form of handwritten notes from that newspaper. The notes themselves may not be exhaustive. For a more detailed look at entombed toads the *Fortean Times* occasional paper *Toad In the Hole* is well worth looking at.

Statement of the collier who found the frog in Trafalgar Colliery on Thursday Jan 14th 1915.

"I am a butty collier working in the Twenty Inch or Smith Coal seam in the No 4 district of the Trafalgar Colliery. At 4 o'clock on the morning of Thursday Jan 14th I was. work at the coal face. The seam was about 20 to 22 inches thick and the holeing which is thin, was underneath the coal. We had holed in some 15inches and was the usual custom to enable me to conveniently hole further underneath, I then stuck the pick into the coalface, about 6 inches from the floor. At the point at which I did, there was a thin line of hard black earth in which we term "mother" running along the seam. The pick, when I (?) the coal, appeared to strike through into a space. The piece of coal below the pick fell away to the ground exposing a small

Trafalgar Colliery. Near Cinderford. 1207.

cavity, out of which at the same time, a live frog fell. It was small in size and dark in colour, with a bright yellow band running down the whole of the back. It hopped about...." Ernest Giles Brain (various letters underneath).

Letter from the Secretary of the Zoological Society 27th Jan 1915:

'As a fellow of the Zoological Society in London, I have been in communication with the Secretary of that society on the subject of the frog found at the Trafalgar Colliery and brought to our notice by Sir Francis Brain. Sir Francis has expressed a wish that those competent to give an opinion would do so through your columns..... Yours faithfully W.Herbert Drummond FZS'

Dean Forest Mercury January 29th 1915. The Trafalgar Frog:

'The frog or the toad, for there is a lively discussion as to exactly what the amphibian should be known, is now dead but had it survived it would have made its way into the annals of the Forest of Dean for many years....It leapt as long as it had its freedom and it leapt to the top of the receptacle in which it was placed when it was being photographed...meantime we may state that the small cavity appears to be a little larger than the size of the frog but the shape shows a good deal of resemblance to the form of the reptile, including a pushing out portion which may have contained its head. The body in which, in what ever way it may have originated, perhaps went there a very long time ago. If this theory is correct why it should be a frog and not a being of only a remote relationship to the perfected amphibian, we will leave to the evolutionists to explain... With just caution from the lay mind, evolution is a mighty slow process and when a frog was not a frog, well it may be difficult to exactly define and definitely determine.'

Dean Forest Mercury February 5th 1915. Shows a photograph of the lump of coal where the frog was found in and the bottle in which the frog is preserved.

Extract from letter dated February 14th 1915:

> 'I have in my possession a frog discovered some 20 years ago in the Starkey Seam at the Duck Colliery about 80 yards below the surface....The collier was holing by removing the solid shale over the coal seam, when in the act of doing this, he liberated the reptile which jumped out of its prison...I have always regretted that the cavity was not preserved to silence all doubt as I was not aware of the find until it was too late for this. The late Mr Arnold Thomas found a frog in an adjoining pit in the same Starkey Seam, some years before the find in the Duck Pit. Yours Faithfully Joseph Hate' (?)

The following is all that is written:

Dean Forest Mercury (page 4) 19th February 1915. More letters about Trafalgar frog (sic). Trafalgar is presumably a town or village in Gloucestershire.
Dean Forest Mercury. p.3. Feb.26th 1915. More on Trafalgar frog.
March 5th 1915. page (?). Letter from a scientist Frank Brain and A.Trigg.
March 26th 1915. Page 6 lower right hand corner more on Trafalgar Frog

`Planet Claire`-*The B52s.*

Planet Claire has pink air
All the trees are red
No one ever dies there
No one has a head...
Some say she`s from Mars
Or one of the seven stars
That shine after 3.30 in the morning

BUT SHE ISN`T

WEDNESDAY, OCTOBER 28, 2009

MUIRHEAD'S MYSTERIES: The Kaiser`s caterpillars - early plague warfare or an U.F.O. scare?

Dear folks,

Today`s blog is about a strange occurrence during World War One near the Jenkins Chapel area of the Peak District, in the vicinity of Saltersford. I wrote a letter to *Animals & Men* a few years ago asking for further information but none turned up. Like so many Fortean stories I have found, this one was discovered whilst I was a patient in a psychiatric hospital. But it wasn't a delusion, because I have the extract right in front of me. The story of the Kaiser's caterpillars was published in *In And Around The Peak District* by Doug Pickford (1993), a Macclesfield historian who has been connected with *Old Macc* magazine.

Pickford asks:

> 'Did a German Zeppelin drop millions of caterpillars around the Jenkin Chapel area of Saltersford? An unlikely if not preposterous question may well say, but some senior Saltersford and Rainow residents are firmly of the opinion that this very much the case during the First World War. Since hearing of this strange incident I have spoken to a number of inhabitants of the area who can recall the events that unfolded during the dark days of war in the year 1917, when a plague of millions upon millions of caterpillars descended on the area.' [1]

Now, when I first read this I thought of three things: I have read about swarms of insects appearing, even in the unlikeliest of places, such as in urban areas. But I have not heard of millions upon millions; that does seem a lot. I am no entomologist, but Pickford describes them as being 'furry' [2] and 'black and yellow', 'some one and a half inches in length' [3] and a great number of crows ('thousands' [4]) came from a wide area to feast on the multitude of insects. Also, I remembered that pre-World War One was a time of phantom airship scares, which I know nothing about. Does anyone know of associations between swarms of insects and U.F.Os? Thirdly, if this wasn't an early U.F.O. was it a German attempt at 'plague warfare'? Note the caterpillars, if that is what they were, appeared just in time for the harvest.

Pickford picks up (if you'll pardon the pun!) the story at around harvest time, 1917. This would be around the time of the Russian Revolution:

> 'The only Zeppelin - a gas filled balloon powered by propellers - known to have flown over that particular area came one moonlit night in 1917. The exact date has not been ascertained but it was around harvest time. The German aircraft is said to have dropped a bomb at Pott Shrigley but it did not explode and then it turned and flew over the valley, over Rainow and on to Saltersford. It was eventually brought down when it reached the coast. However, that night and the following morning locals discovered literally millions of black and yellow coloured furry caterpillars some one and a half inches in length. They were everywhere. The plague of wriggling creatures appeared to be centred on Greenstacks Farm where all the downstairs rooms were covered inches high with the creatures. Green Booth and Hollowcowhey Farms were also affected very badly. My 83 year old informant told me "It was though a stone had been thrown in a pool, with the ripples strong in the middle at Greenstacks and they went out for about a mile in circumference. Farms were almost bankrupt after. The caterpillars had eaten everything. There was no grass, no greenery at all growing. There was no food for the cattle and there was no hay to be harvested. Afterwards the area was black where grass and crops should have been [this blackness is interesting, it occurs in some mystery

snake reports-RM]....And then the crows came. Apparently "thousands of crows" came from all around and started to eat those furry caterpillars with a vengeance. They gorged themselves until they were so full they could not fly. A lot of them managed to get on to the tops of the dry stone walls and stayed there for hours, unable to move off. The walls were turned white with their droppings.' [5]

The story goes on to describe how the curate of St Peter`s in Macclesfield [this may be the same as the church I used to go to-RM] was posted to Rainow during the Great War. He collected some of the caterpillars but there is no record of what he found out about them. As for the villagers: '...It was a frightening experience for them, not least because they appeared from nowhere...dropped from the sky, perhaps, as a secret weapon by an enemy hell bent on destruction? Or was it one of nature`s aberrations?' [6]

A Jenkin was a drover who preached at a 'preaching cross' at a crossroads in this area.

1. D. Pickford. *In And Around The Peak District* (1993). pp5-6
2. D. Pickford Ibid. p7
3. D. Pickford Ibid.p7
4. D. Pickford Ibid. p.8
5. D. Pickford Ibid pp 7-8
6. D. Pickford Ibid p 10.

Muirhead`s Mysteries will be taking a short break till Tuesday November 3rd. I will be 43 on the 5th. Yipee, bang, wizzzz!! - Bonfire Night. Rare crypto books and bottles of ginger wine kindly appreciated. Just don`t ask me to blow up Parliament, I'm not that radical!

> October, trees are stripped bare, of all they wear,
> Do I care?
> October, kingdoms rise and kingdoms fall,
> But You go on and on. (*U2* `October`)

WEDNESDAY, NOVEMBER 04, 2009

MUIRHEAD'S MYSTERIES: The wonderful Battell of Starelings....Cork, Ireland, 1621

Hello again folks, Muirhead`s Mysteries is back again in a blaze of glory, or feathers as the case may be:

The Journal of the Cork Historical And Archaeological Society Vol 2A 1893, William Corliss`s *Biological Anomalies* Birds, (1998) John Michell and Robert Rickard`s *Living Wonders* (1983) all report on a battle amongst starlings in Cork, Ireland in 1621. This was originally recounted in a pamphlet held at the British Library (in 1893) that is, not today necessarily titled: `The Wonderful Battell of Starelings Fought at the City of Corke *in* Ireland, *the 12*[th] *and 14*[th] *of October last past,1621*. This was published in 1622.

"About the seventh of October last, Anno 1621,there gathered together by degrees an unusual multitude of birds called stares, in some Countries knowne by the by the name of starelings. These birds are for the quantity of their bodies strong, for their quality bold and ventrous, among themselves very loving, as may appear by their flights keeping together all times of the yeare, excepting breeding time. ...The stares, or starlings, they mustered together together at Cork some foure or five daies before they fought their battells, every day more and more increasing their armies with greater supplies..." [1] [Corliss, Michell and Rickard take up the story in *Living Wonders* [2] :]

Prior to the famed battle, two large flocks of starlings had been massing, one to the east of Cork, the other to the west. The birds were kept within the bounds of their apparently well-defined territories until 9:00am on October 12th 1621. Bourne described what happened next:

"Upon a strange sound and noise made as well on one side as on the other, they forthwith at one instant took wing, and so mounting up into the skies, encountered one another with such a terrible shock, as the sound amazed the whole citie....Upon this sudden and fierce encounter, there fell down into the citie, and into the rivers, multitudes of starelings, some with wings broken, some with legs and necks broken, some with eyes picked out, some their bills thrust into the breasts and sides of their adversaries, in so strange a manner, that it were incredible except it were confirmed by letters of credit and by eyewitnesses, with that assurance which is without all exception.

The conflict dragged on for two more days. On the 13th, the carnage took place at some distance from Cork. The battle on the 14th was again over Corke, and dead birds again rained down on the city." [3]

Corliss then go on to detail a battle between ravens over Ginnhein, Germany in 1883:

"*The Frankfurt* (Germany) *Journal* writes: The gardener, Mr Georgius from Ginnhein, called at our office today with a chest full of dead ravens, victims of a battle which was fought high in the air among a flock of over four hundred of these birds near the above mentioned village. The ravens formed together into three detachments, and as if at a given signal flew at each other, and with savage cries seemed as if they would tear each other's eyes out of their heads with their beaks". [4]

Corliss has this to say about bird battles:

"Anomaly Evaluation: We can only surmise that two bird armies would clash so viciously only because of a territorial dispute involving, say, choice nesting and feeding sites. The anomalous aspect of such battles is seen in the sharp division of the forces, their organisation, communication, discipline, and perhaps even strategy. Such characteristics are rarely seen in vertrebates, except, of course, among some primates!

Possible Explanations. Territorial disputes.

Similar and Related Phenomena. Some insects, particularly ants, engage in battles, but usually with different species." [5]

There are probably many other records of bird battles buried in the literature. On page 393 of *The Rough Guide to Unexplained Phenomena* also by Michell and Rickard 2ⁿᵈ ed. There is an illustration from 16ᵗʰ century India of a battle between owls and crows.

1.   J.C. The Wonderful Battell of Starelings Fought at the City of Corke, in Ireland, the 12ᵗʰ and 14ᵗʰ of October last past,1621 *Journal of the Cork Historical and Archaeological Society* Vol 2A 1893 p.260-261
2.   J. Michell and R. Rickard *Living Wonders* (1983) p.154 in W, Corliss *Biological Anomalies: Birds* (1998) p.246
3.   W. Corliss *Biological Anomalies: Birds* (1998) pp246
4.   Corliss Ibid. p.246
5.   Corliss Ibid p.246

That`s all for today folks. Supposing YOU the reader of this blog suggest something I can blog about and I`ll see if I have the data?

"Come in Boogi Boy, you`re late! Have you got the papers China-man gave you?" "Yes,Boogi,in the past this information has been suppressed, but now it can be told, every man, woman and mutant on this planet shall know the truth about De-Evolution." (*Devo*-Jocko Homo, video version. Do me a favour, for my birthday on the 5ᵗʰ, watch this video on the Net.)

Richy

THURSDAY, NOVEMBER 05, 2009

MUIRHEAD'S MYSTERIES: An attack upon a 19 year old girl by a yeti in Nepal in 1974

Hello folks,

I was browsing along my bookshelves yesterday looking for something to talk about today when I came across a paperback: *The Hell Hound and Other True Mysteries* by Peter Haining. In it was a story that caught my eye: 'The Girl Who Fought An Abominable Snowman.' So today I am going to summarise this case. I have not had a thorough look through the rest of my crypto books so my apologies if this is well known. Even if it is there may be some of you out there who are unfamiliar with it. Nor am I aware of any other reports of mystery hominids attacking humans. Richard F – do you know any accounts from amongst the almas and orang pendek?

> 'This is the story of Lakpa Sherpani, 19. She lived in the Khonar district in Nepal as a yak herder in the shadow of Mount Everest and had done this sort of work since a child. But in 1974 life was for a brief moment to become anything but ordinary for Lakpa. Indeed it was to become so extraordinary that she made headlines around the world and added a new chapter in the story of one of our great mysteries.' [1]

The author (Peter Haining) then goes on to describe her interest in Nepali folklore:

> 'Indeed it was her knowledge of folklore that was to help her to be so precise about what happened to her on that bright summer morning in 1974. On this particular day, Lakpa was watching over the herd as they rested half a mile or so from one of the dense forests which are found throughout the Himalayan region. These forests are not spots to go for grazing for they are invariably dense with vegetation, mostly fog-enshrouded and always inhospitable. Many a local legend surrounded them, and only a foolish or woman would enter one....'[2].

The author then describes, in a possible embellishment of the drama, how the young woman felt movement near her and the presence of a bulky body:

> 'Next, a hand grabbed at her hair and forced her head back. She felt a body straddle her own, and was suddenly conscious of the most overpowering stench. [I think I`m right in saying that this is a feature of some mystery hominid reports - R.M.] When she opened her eyes she found herself looking at a face that was hardly human – or animal. Tiny yellow eyes glared at her from a face covered with hair. Prominent teeth protruded from a huge, slack jaw and the whole face was surrounded by long brownish hair. Lakpa screamed. Another hand, which she could also feel was covered with hair, struck her across the face as if in response, and she suddenly tasted her own blood on her lips. ...Her senses had already told her this was no human being. It was not much bigger than a child - perhaps four feet tall - but with many times the strength of an infant. Through her agony she also saw that the creature was naked except for a complete covering of hair, brownish in colour to the waist and black and thick on the lower stomach and legs.'[3]

The author then describes how the creature knocked her unconscious with a blow to her head and then made an attack on her yaks.

> 'The young girl licked the blood from her lips and watched what the creature was doing. It was running amongst the yaks, seizing hold of some of them and twisting their horns until they fell to the ground. Already a couple were down and showed no signs of movement, and it was evident the creature had enormous strength.' [4]

The story then goes on to describe how Sherpani has heard about attacks by yetis on yaks. And:

> 'One of the reasons for such universal interest in Lakpa`s story, was that a group of three British zoologists had just published a report of a study they had made in the Himalayas on the likelihood of there being any such creature as The Abominable Snowman. The men, Mr J. A. McNeely, Mr A.W. Cronin and Mr H. B. Emery, three experienced and highly qualified observers, had not only confirmed the possibility, but stated that during their stay in the area, one of the creatures had actually visited their tents and "left tracks that are not referable to any known animal"' [5]

It would be interesting to track down this study. These zoologists apparently thought that the yeti is Gigantopithecus. BBC Online`s website also carried a very brief report about Lakpa`s encounter on its So Weird, Lionel`s Guide The Ape Type site.

Tomorrow – a report on Fawke`s pet winged cat, Fluffy, the real ring leader behind the gunpowder plot. Er, some mistake there! Would have made a great blog if it were true, though!

1.   P. Haining *The Hell Hound and Other True Mysteries.* (1984) p.81
2.   P. Haining Ibid p.83
3.   P. Haining Ibid pp84-85
4.   P. Haining Ibid p.85
5.   P. Haining Ibid p.89

When I looked out the window
On the hardship that had struck I saw the seven phials open
The plague claimed man and son

Four men at a grave in silence with hats bowed down in grace
A simple wooden cross,
It had no epitaph engraved..
It had no epitaph engraved

Come on down and meet your Maker
Come on down and make the stand
Come on down, come on down
Come on and make the stand. (*The Alarm`* The Stand`)

FRIDAY, NOVEMBER 06, 2009

MUIRHEAD'S MYSTERIES: The Limerick Cathedral misericords

Dear folks,

Today I take a look at the misericords of Limerick Cathedral. A misericord is a:

1. Ledge projecting the underside of a hinged seat in a choir stall, giving support to someone standing when the seat is folded up.

2. An apartment in a monastery in which some relaxations of discipline were permitted.

3 A small dagger for delivering a death stroke. [1]

Seeing as I don`t want to become a monk or assassinate the Supreme Leader of the CFZ, for the purposes of this blog I will be referring to definition 1.

Historian John Hunt has commented:

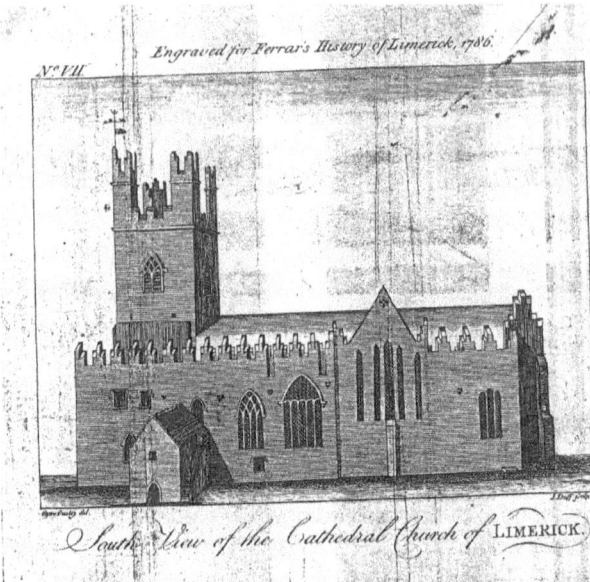

Engraved for Ferrar's History of Limerick, 1786.

N°VII

South View of the Cathedral Church of LIMERICK.

'In the Middle Ages, cathedrals and churches presented a very different appearance to that which we now see. Until the sixteenth century changes, and before the destruction and desecration which took place under puritan hands in the seventeenth century, the great Irish cathedrals vied with those of the Continent and England in the richness and beauty of their interior decoration and carved woodwork and furnishings. Walls were covered with paintings, and elaborate screens and portions of carved tracery work marked the divisions of the church into nave, chancel and side chapels. ..There are now only nineteen misericords remaining on the stalls in Limerick cathedral. Their arrangement has been altered several times during the last century and at present there is only one range of stalls on either side of the chancel in the cathedral. Originally there were probably two, together with seats or forms for the boys of the choir...The creatures carved on the misericords come from that wonderfully rich world created by the medieval imagination out of half understood and oft repeated travellers` tales. Bestiaries, books containing accounts and illustrations of beasts fabulous or otherwise, were very popular throughout Europe from the twelfth century onward. These became the pattern books from which the medieval sculptor drew much of his inspiration

Like most things in medieval life, each beast usually had a mystical or religious significance. The mind of medieval man seized upon these creatures, and saw in each a secondary significance often pointing some moral or religious lesson underlying their ordinary or extraordinary appearance. So every carving carried a message more widely understood in the fifteenth century than it is today.' [2]

Some examples of the Limerick Cathedral misericords are as follows.

'A griffin. The body and limbs are those of a lion with the wings and head of an eagle. It is immensely strong and tears in pieces men and horses which it especially hates.' [3]

'The manticora. He inhabits the Indies. He has the head of a man, the body of a lion, the wings of an eagle, the tail of a scorpion, and feeds on human flesh.' [4]

'A yale (?) Often in heraldry, this was an animal like a horse with long with long moveable horns and reputed to have the power of killing with a glance, or perhaps a unicorn, an image of Christ.' [5]

'An amphisbaena. This curious beast is like a wyvern, but has an additional head at the end of its tail. Evil can proceed in more than one direction.' [6]

There are at least two wyverns ('a wyvern is a winged two legged dragon with a

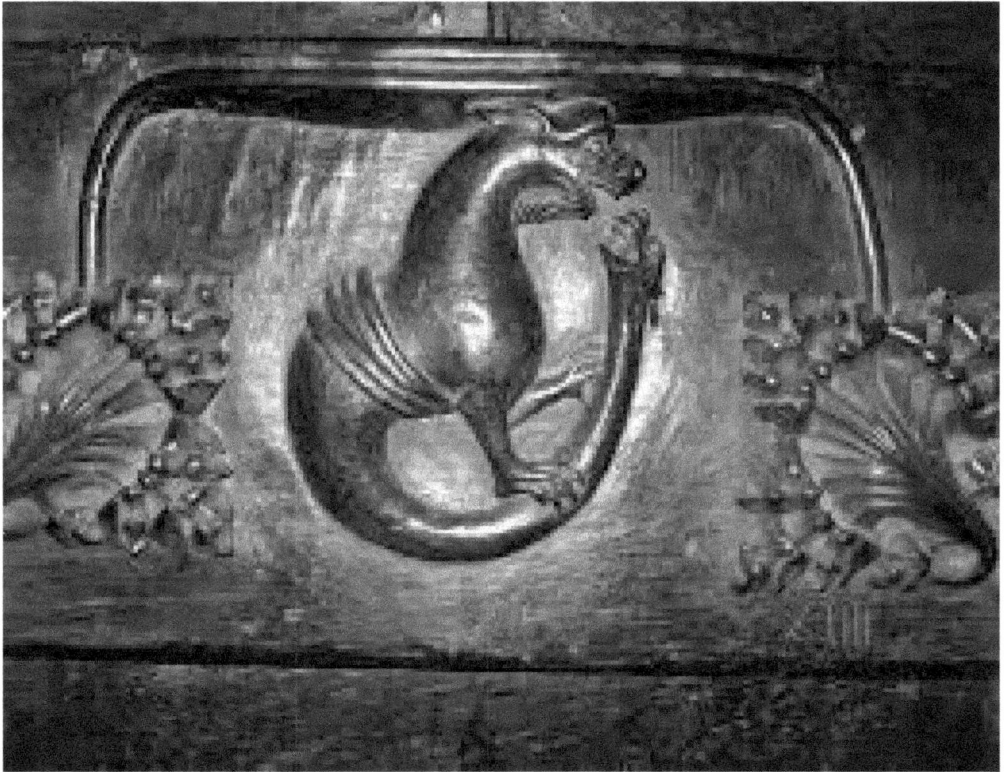

barbed tail.' [7] amongst the misericords. One with a head curved back biting its tail and another with a raised head and curved tail.

There are several sites on the Web showing good quality illustrations of the Limerick Cathedral misericords; just Google `Limerick Cathedral Misericords.' Also, at http://www.misericords.co.uk/bibliography.html there is a comprehensive bibliography on misericords.

1. *Concise Oxford English Dictionary* (2008) p913
2. J.Hunt The Limerick Cathedral Misericords. *Ireland of The Welcomes. Vol.20* (3) 9-10. 1973. pp12-13
3. J.Hunt Ibid p.13

4.   J.Hunt Ibid. p.13
5.   J.Hunt Ibid .p.15
6.   J.Hunt Ibid. p.16
7.   *Concise Oxford English Dictionary* (2008) p.1667

Thanks for the poem on the blog for my birthday Jon and Richard. I was going to quote from Stevie Wonder's song *Happy Birthday* here but decided it would be hubristic. The song is for Martin Luther King, so then I thought of quoting from `Belfast Child` by Simple Minds.

> Brother, sister where are you now
> As I look for you right through the crowd
> All my life here I`ve spent
> With my faith in God our church and the government
> Some say troubles are bound
> Some day soon they`re gonna pull the old town down
> One day we`ll return here
> When the Belfast child sings again
> When the Belfast child sings again.

SATURDAY, NOVEMBER 07, 2009

MUIRHEAD'S MYSTERIES: Interesting insects part one

Dear folks,

Today I am commencing part one of a multi-part series titled Interesting Insects. (OK, I know that scorpions, which feature in one of today's stories, aren't insects, but interesting creepy crawlies would spoil the alliteration). These stories will be gleaned from a number of sources, such as web sites, newspaper cuttings, and letters from other Forteans or cryptozoologists. As is mostly my practice I will be presenting the information in the order in which the occurrences or instances of unusual insect life take place, which is not necessarily the same date as the publication of the book. For example, if a book published in 2009 reports a strange moth in Birmingham in 1909, then the latter date is the one to look out for. Parts one to three are based on observations I have collected in the sources mentioned above whilst later, observations by our mentor Charles Fort were published in his complete books.

Hopefully an issue of *The Amateur Naturalist* in 2010 will be publishing an account of my trip to Hungary in May 2009 to study the fauna of the Aggtelek National Park, particularly its butterflies.

Inevitably most of my cuttings date from the last 35 years or so, this being how long I have been active in cryptozoology/Forteana.
So here we go. The earliest record I have worth recounting is from the 1600s, from antiquarian

John Aubrey`s *The Natural History of Wiltshire* (1847 edition, originally published in c. 1691), my copy of which a colleague of mine at Thornton`s Bookshop in Oxford bought me, before the bookshop became defunct:

'Riding in the north lane of Broad Chalke in the harvest time in the twy-light, or scarce that, a point of light, by the hedge, expanded itselfe into a globe of about three inches diameter, or neer four, as boies blow bubbles with soape. It continued but while one could say one, two, three, or four at the most. It was about a foot from my horse`s eie; and it made him turn his head quick aside from it. It was a pale light as that of a glow-worme: it may be this is that which they call a blast or blight in the country.' [1]. Broad-Chalke is the next village westwards of Bishopstone, near Salisbury where I spent that part of my childhood that wasn`t spent in Hong Kong.

Now the interesting thing is that Aubrey compares this to a 'glow-worme'; he does not say it *is* a glow-worm.

Concerning the Giant Earwig of St. Helena *(Labidura herculeana)*: 'This extraordinary insect, which reaches the huge (for an earwig) size of between $2^1/_2$ and 3 inches, was discovered in 1798 by the Danish zoologist, Fabricus. It then vanished for nearly two centuries. In 1962 an expedition looking for bird bones buried in the sand of this isolated and tiny South Atlantic island discovered body parts which appeared to come from a giant earwig. Three years later another expedition found living ones, but since then no Giant Earwigs have been found. It seems almost certain, however that, in this particular case the truth IS out there' [2] Five years prior to this CFZ report, *The Guardian* reported:

> 'They [Paul Pearce-Kelly, then a zoologist at London Zoo and Dr Graham Drucker a zoologist at the World Conservation Monitoring Unit, Cambridge] are...in pursuit of a giant beetle – also believed to be extinct – and three species of blushing snail, of enormous importance to evolutionary theorists. [3]

Jumping ahead to 1870, here is a note from a journal called *The Naturalists Note-Book* on a favourite of mine, British locusts: 'Locust-On the 11th November, in a bright gleam of sunshine

> We caught a locust on the jessamine growing over the front of our house. Thinking it more merciful to have it quickly despatched, and wishing to preserve it intact, I applied to a chemist, who speedily destroyed it with prussic acid, that fact that it was feeble rendering it easy to adminster the dose...I presume it is a migratory locust; but how is it that there is no green colour anywhere? There is a little bit of red on the body; but generally it is of a flat, grey hue. It is about two and a half inches in length; that is, the body without the legs; and the spread of the wings from tip to tip, nearly five inches. I hear

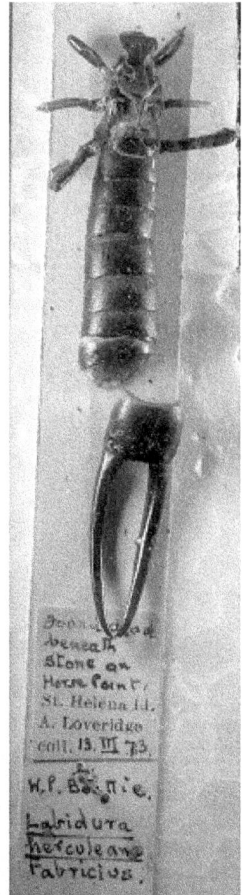

that several have been caught in this neighbourhood during the last few weeks. How do they get here? Are they stragglers blown out of their course? Or is it possible that they are brought in some cargoes? I have had the above account sent to me by my sister, who resides at Truro, in Cornwall.-H.BUDGE' [4]

Finally, a mystery that dates from at least 1872:

'A colony of scorpions is alive and well and living on the Isle of Sheppey. The cold British winters should have seen off the scorpions years ago, yet the colony of *Euscorpius flavicaudis* has survived for more than 120 years-although probably not in great comfort...With no natural enemies on Sheppey apart from the occasional human who might tread on one, these scorpions have survived because of their remarkable adaptability. They retreat deep into their cracks to escape the cold and do nothing but wait for the next woodlouse.' [5]

1.   J. Aubrey *The Natural History of Wiltshire* 1847 p.18
2.   J. Downes. *Mystery Insects* C.F.Z website. Downloaded April 15th 1998
3.   T. Radford *Scientists set off on the track of the giant earwig. The Guardia*n. September 23rd 1993.
4.   H. Budge Locust. *The Naturalist`s Note-Book* 1870 p.44
5.   Anon. Scorpions in a cold climate. *New Scientist* May 16th 1992. p.15

Richard will be back on Tuesday.

For any of you going through hard times tonight...

U2- `Drowning Man`:

Take My hand,
You know I`ll be there,
If you can I`ll cross the sky
For your love,
And I understand
These winds and tides,
This change of times
Won`t drag you away.

Hold on, hold on tightly,
Hold on and don`t let go
Of My Love
The storms will pass
It wont be long now,
His love will last
His love will last, forever.

TUESDAY, NOVEMBER 10, 2009

MUIRHEAD'S MYSTERIES: Interesting Inverts part 2

Hi folks,

Part Two of my blog includes mention of spiders, which of course are not insects. I start with an obscure reference from 1904, in the form of a communication in *Notes and Queries*:

> 'As regards the venomous spider in China, I too have heard that there is one, but as to its name I cannot speak.-Yours very truly, John Batchelor. Let us hope that some society, or some wealthy friend of learning and of missionary civilization, will find the funds for publishing Mr Batchelor`s laborious work before he dies. I had told him that there is in New Zealand a venomous spider called katipo by the Maoris, and that there is said to be another in China bearing the same name in Chinese. Is that a fact? The Religious Tract Society, 4, Bouverie Street, E. C. has lately published `The Ainu and their Folk-lore.` E.S.Dodgson [1]

Fortean Times no. 242 November 2008 carried a feature titled `The Spider Monster of Issoir` by Theo Paijmans - a tale of giant spiders in nineteenth-century Paris. American newspapers such as *The Sandusky Register* for February 1st 1895 reported, on the death of a Parisian:

> 'The countryman was lying on his back writhing in the grasp of an unknown monster, whose horrible aspect froze the agents of police with terror. It was as large as a full-grown terrier, covered with wart like protuberances and bristling with coarse brownish hair. Eight jointed legs, terminated by formidable claws, were buried in the body of the unfortunate victim. The face had already disappeared. Nothing could be seen but the top of the head, and the monster was now engaged in tearing and sucking the blood from his throat.' [2]

This reminds me a bit about the scene with the priest Nathanial and the blood-sucking Martian in H.G. Wells`s *War of The Worlds*.

Jumping ahead to 1986 and the Chernobyl nuclear disaster in the Ukraine in April of that year, this report appeared in *The Times* on February 14th 1987:

> 'Bees knew secret of Chernobyl disaster: Polish bees headed straight back to their hives when they sensed contamination from the Soviet Union`s Chernobyl nuclear disaster while the rest of Poland was still in the dark about the accident, according to a beekeeping expert. Mr Henryk Ostach, who heads the Polish beekeepers` association, said apiarists were baffled when bees hid for several days after the explosion at the reactor. `When the explosion occurred, the bees interrupted their flight, although it was a fine sunny day. Not yet knowing anything about what had happened at Chernobyl, we wondered why the bees suddenly hid in their hives. They surrounded their queen very closely, beating their wings constantly in order to minimize the permeation of contamination.` [3]

Moving ahead to February 1992, we had an occurrence of a Mole cricket in the South Pacific. This was passed on to me by Darren Naish:

> '8 February 1992. AT 2000 UTC whilst approaching the coast of New Zealand a beetle or grasshopper-type insect, was found on the starboard bridge wing, apparently basking in the early morning sun. When approached it reared up on its back legs and started to move its front legs in a grasping motion. The front legs appeared to be developed into `armour-plated` gripping devices as they had hooks and claws along the lower and front edges. The head was also

MOLE-CRICKET, WITH EGGS AND LARVÆ (slightly enlarged).

armour-plated with four tube-like antennae protruding from the front of the head beside the jaws. Set above these were two feelers which were 23mm long and the insect had powerful jaws. [4.] Darren wrote besides the artists impression of the mole cricket: "I think this is a mole cricket...but I didn't know they lived in the South Pacific."'

Lastly, coming right up to date with a report from *The Guardian* November 5th 2009:

Towering ants nests in woodland get protection:

"A rare "skyscraper city" made by ants has been given the equivalent of listed building protection and a place on maps to safeguard it from forestry work. Nests up to two metres (7ft) high, constructed from conifer needles in Northumberland woodland, will be monitored during the felling of "intrusive" 20th century conifers in Holystone, near Rothbury. The whereabouts of 69 structures, made by colonies of the hairy northern wood ant, have been plotted. The species is Britain's largest, but on a human scale, the nests dwarf the ants by a greater measure than the Empire State Building". [5)]

That's all, folks. For reasons too tedious to go into I cannot provide song lyrics today but they will be back tomorrow.

1. E. S. Dodgson. Untitled article. *Notes and Queries* April 2nd 1904 p.265
2. The Sandusky Register in T. Paijmans The Spider Monster of Issoir *Fortean Times* November 2008
3. Anon *The Times* February 14th 1987. Bees knew secret of Chernobyl disaster. Anon Insects. South Pacific Ocean.
4. *The Marine Observer*. 63(319) Jan.1993 pp14-15. *The Marine Observer* occasionally reports on unknown animals or surprise occurrences.
5. M. Wainwright. *The Guardian* November 5th 2009. Towering ants nests in woodland get protection.

WEDNESDAY, NOVEMBER 11, 2009

MUIRHEAD'S MYSTERIES: Interesting Invertebrates Part Three

Hello again,

Today I am going to look at some cases involving insects and spiders from various parts of the world, mainly clustered around the late 1980s and early 1990s. There is no particular reason for this; it is just that I happen to have quite a large collection of cuttings from this time period.

This extract is from *BBC Wildlife Magazine* July 1984.:

'BUZZ: The world's biggest bee has been rediscovered in the rainforests of Indonesia. The species, *Chalicodoma pluto*, was presumed extinct, having not been seen since it was discovered in 1859 by Alfred Russel Wallace. He collected the only specimens known to science (two females) on the North Molluccan island of Bucan, and that was the extent of our knowledge until an American biologist, working on the neighbouring island of Halmahera in 1981, heard an ominous buzzing...." The bees built communal nests, which 'were always built inside those of tree-dwelling termites, though the resin and wood-fibre galleries constructed by the female bees were tough enough to resist termite invasions. The huge mandibles of the female, reminiscent of a male stag beetle, were found to be tools for scraping resin, which was then bulldozed by the central labrum into a 10mm diameter ball....' The article then goes on to describe how the male of the species are much smaller.' [1]

'Locust retribution for Washington's sins', *Guardian* May 5th 1987:

After biding their time for 17 years buried deep below ground in the luxuriant spring gardens of Washington, a plague of locusts is preparing to descend upon the capital. The curious creatures, which have sent Washingtonians to the hardware stores to block up their windows and doors with defensive screens, have been quietly awaiting their moment to attack...Many millions of creatures have already tunnelled their way to the surface and are waiting the right moment to take part in one of nature's most unusual dramas. Those people who have lived through it before – in 1902, 1919, 1953 and most recently, in 1970 – say it is an experience they will never forget. Local garden stores warn there is nothing you can do. [2]

So if this number 17 was consistent, then 2004 should have been another significant year. But I have no records from 2004. The next plague year is 2021. Michael Oliver wrote to *The Guardian* on May 9th 1987:

'A plague on their houses: Sir, - The article by Alex Brummer (May 5th) is quite intriguing. It seems that this plague of locusts in Sodom recurs in 17 year cycles – 1902, 1919...1953, 1970, 1987. The interval between 1919 and 1953 is 34 years, i.e. 2 x 17, and it seems therefore that 1936 was for some reason 'spared.' However, this multiple factor of 17 would appear to be too much of a coincidence, and I am wondering if there is a scientific explanation for it.' [3]

A letter dated May 13th 1987 from L. E. Mack stated that:

'[Richard] Dawkins claims that the only suggested explanation of this so far offered is that the numbers 13 and 17 are prime numbers. The advantage of regular 'plague' eruptions is that the insects can alternately "swamp" their predators: if these eruptions are timed so that they occur at intervals a prime number of years apart, predators do not get the chance to 'synchronise' their own cycles. So when the food is there, they are not. [4]

Now, the strange case of the spiders' webs in the Croatia-Serbia war of 1991. On November 8th 1991 'Zagreb radio claimed "large quantities" of live, yellow-backed spiders were dropped on the east Croatian town of Daruvar. "The spiders are being used by the aggressor army as a biological agent". On television, an unnamed doctor said there was no trace of poison among the samples he had seen. "But the next generation could be the killer strain," he cheerily confided. In fact, the tale of Serb-trained killer arachnids ranks as one of the sillier fabrications in an alarmist propaganda war, fit to stand alongside earlier claims that Croats in the town of Osijek had freed tigers from the local zoo which then roamed the countryside lunching on Chetnik guerrillas.' [5]

Finally, a headline from the *Cork Examiner* September 26th 1994: 'Extinct' insects discovered near Thurles: 'Very rare insects thought to have been extinct in Britain and Ireland have been discovered near Thurles. Now zoologists and entomologists from both sides of the Irish Sea are putting the insects under the microscope. "The discovery is very exciting and a big breakthrough" said Tom Grace, Chairman of Cabragh Wetlands Trust near where the insects were found. The rarest insect is called *Limnephilus pati*, a caddis fly resembling a moth. Two males and two females of the species were found.' [6]

1.   Anon. Buzz. *BBC Wildlife Magazine*. July 1984.
2.   A. Brummer. Locust retribution for Washington's sins. *The Guardian* May 5th 1987.
3.   M. Oliver. A plague on their houses. Letter in *The Guardian*. May 9th 1987
4.   L.E. Mack 'Nymphs and shepherds of Sodom-on-Potomac' Letter in *The*

Guardian. May 13th 1987
5. I. Traynor. Croatian media at battle stations for attack by eight-legged yellow peril. *The Guardian* November 12th 1991.
6. Anon. `Extinct` insects discovered near Thurles`. *Cork Examiner* September 26th 1994

Sorry, I forgot to look up a song lyric again. But try listening to `Batchelor`s Hall` by Steeleye Span.

THURSDAY, NOVEMBER 12, 2009

MUIRHEAD'S MYSTERIES: Interesting invertebrates part four

Hello again,

Here is another collection of stories of a Fortean nature concerning invertebrates; both spiders and insects. The first story is from *The Guardian* of May 30th 1987 `Death in Venice for marauding gnats.`

The article begins by mentioning the forthcoming summit of world leaders in Venice. Then,

> 'But for the leaders` Venetian hosts, one less publicised problem is multiplying daily by the million; the gnats of the Lagoon are beginning to swarm. Last summer, dense aerial formations twice shut down Venice airport; frequently, the waterbuses chug to a halt in the face of the ubiquitous insects. The fear is that the summit, to take place on the island of San Giorgio Maggiore, could be blotted out, hidden from sight and even worse, from television cameras...Venetians insist that the gnats do not bite, just get in the way. But even so, no chances are being taken with the leaders. At Venice city hospital, two private rooms are being made ready for any illustrious patient – and all staff leave has been cancelled.' [1]

Now moving on to spiders; a rich source of lore. Unknown paper (possibly *The Daily Telegraph*) August 6th 1991:

> `Blind spiders are a sight for sore eyes`: A colony of blind spiders previously thought to be extinct has been found in a remote cave on the Nullarbor Plain in South Australia. Known as *Troglodiplura Lowryi*, the spiders are related to the trapdoor and funnelweb species. Only two other types of eyeless spider are believed to exist. Their straw-coloured bodies are about one inch across. They lost their pigment and their eyes thousands of years ago when the species moved hundreds of feet underground and more than a mile from the cave entrance. A research team from the Australian Museum`s arachnology department did not find any adult spiders but the evidence that the colony was doing well. Only

carcases of the species had been found in the past. [2]

The second spider story, coincidentally also connected to Australia, dates from September 12th 1991, also in *The Daily Telegraph.* `Killer spiders gassed in Norfolk barn.` :

'A barn in which killer spiders were found was sealed off while health officials destroyed the insects with poison gas. The Australian spiders were discovered in the barn in Seething, Norfolk, where a couple had stored furniture from Adelaide. For 66 hours, health workers pumped gas into the barn to destroy the small and mainly black redbacks, which can kill humans. The venom paralyses the respiratory system after two to three days of pain and nausea...Fumigation specialists pumped methyl bromide inside and around the furniture and health officers recovered two dead redbacks...Mr Paul Hillyard, a spider expert at the Natural History Museum, London, said redbacks posed a greater danger to children and the elderly than to healthy young adults. "Deaths from redbacks were common in Australia until about 30 years ago when an anti-venom vaccine was introduced widely," he said.' [3]

Moving on to ants, or rather giant ants, this again from *The Daily Telegraph*, (I've not suddenly turned Conservative by the way, I still intend to keep the Red Flag flying over this corner of Macclesfield some of you will be glad to hear, until that is banished by right-wing hordes, it's just that for a while at least this particular broadsheet was strong on off beat stories) of October 4th 1991:

`Man dies as giant ants swarm into motel room`. A man died after being attacked by a swarm of the world's largest ants, it was claimed yesterday. Mrs Hazel Murphey said she woke to find hundreds of Brazilian fire ants, each as big as a thumbnail, crawling over their bed in a motel room in Houston, Texas. Mrs Murphy, whose husband died later, is suing the motel owners...They are currently threatening construction of the world's biggest scientific instrument, a 17-mile wide atom smashing machine near Waxahachie, Texas [Ah - is this the same thing that has been trying to smash atoms in Europe over the last few months with repeated failure that some commentators are quite seriously putting down to Divine intervention? R] Attempts to kill the ants off have failed because, unlike many insect species, they have several queens per mound, and scientists now fear they will attack thousands of miles of electric cables that will service the machine's twin 53 mile underground tunnels, where atomic particles will smash into each other at almost the speed of light".' [4]

So we have two stories here - ants kill human. God (?) uses ants ...to kill machine!

Finally, the humble wart-biter cricket and I assure you it didn't stop a new runway or shopping centre being built. This from *National Trust Magazine* Autumn 1995:

'The almost extinct wart-biter cricket has been rediscovered at a National Trust property in Wiltshire. Thirteen wart-biters were found at the site which has recently been transformed by a down land restoration programme. The wart biter is very particular about where it lives and, with the help of English Nature, the National Trust is now

managing the estate to safeguard its future. It gets its name from the old Swedish custom of allowing insects to Bite off warts' [5]. (This is an interesting bit of folklore. Anyone care to investigate?)

1. G. Armstrong and S. Tisdall. Death in Venice for marauding gnats. *The Guardian* May 30th 1987
2. Sydney Correspondent. Blind spiders are a sight for sore eyes. *The Daily Telegraph* (?) August 6th 1991
3 G. Bartlett. Killer Spiders gassed in Norfolk barn. *The Daily Telegraph* September 12th 1991
4 A. Berry. Man dies as giant ants swarm into motel room. *The Daily Telegraph*. October 4th 1991
5 Anon. Warts and all. *National Trust Magazine*. Autumn. Autumn 1995.
It's nice here with a view of the trees

eating with a spoon?
They don`t give you knives?
`Spect you watch those trees
Blowing in the breeze
We want to see you lead a normal life.

Peter Gabriel -`Lead A Normal Life`.

FRIDAY, NOVEMBER 13, 2009

MUIRHEAD'S MYSTERIES: Interesting Inverts part 5

Hi again, folks,

Today is the last day I will be using my collection of newspaper and magazine cuttings for a Fortean look at insect and spider invertebrates. Tomorrow I will conclude with Part Six: Charles Fort and invertebrate anomalies. Then on Sunday I will make a divergence to Wild Men in Madagascar.

You may be pleased to know that I am not starting off with spiders today, though they will feature in the blog. Instead: earwigs!

From *Science Gossip* September 1st 1865:

> 'WHITE EARWIG - The other day, I found among some gooseberries, a perfectly white earwig, the eyes being black. I have preserved it in spirits; thinking it very rare. I thought I should like to know whether it is so or not, and whether any of the readers of GOSSIP have met with anything of the kind. − R.F.M. [They are occasionally met with.- ED.SC.G]' [1]

A year or two ago the *CFZ Yearbook* published Part 1 of a collection of Forteana from *Science Gossip* I put together.

Now making a large jump to 1995 and our favourite *The Daily Telegraph* for January 20th 1995: a story relating to my home county of Cheshire.

> `Entomologist finds Britain`s 640th spider`: A spider unknown in Britain has been discovered by scientists in a carpet of quivering moss on a Cheshire bog, Liverpool Museum reported yesterday, writes Roger Highfield, Science Editor. The eight-millimetre black spider lives at Wybunbury Moss, a 15ft-thick layer of vegetation that supports trees, bushes and unique wildlife above a 40ft-deep pool of water. *Gnaphosa nigerrima* was identified by entomologist Mr Chris Felton. The Natural History Museum`s spider specialist, Mr Paul Hillyard said it was "a unique find". There are now 640 species of British spiders. [2]

According to *The New Scientist* for April 29th 1995:

> `Spiders on speed get weaving`. (This reported on the interesting observations by scientists at NASA`s Marshall Space Center in Alabama who doped spiders with marijuana, benzedrine, caffeine and chloral hydrate to see how it would affect the construction and appearance of a spider`s web.) 'Spiders on marijuana are so laid back, they weave just so much of their webs and then...well, it just doesn`t seem to matter any more. On the soporific drug chloral hydrate, they drop off before they even get started... The more toxic the chemical, the more deformed the web. NASA researchers think that with help from a computer program they can quantify this effect to produce an accurate test for toxicity.' [3]

In August 1997 I had a strange correspondence with Terry Hooper on an unidentified insect in Ipswich, Suffolk:

> 'Re. insects. A friend in Ipswich has been talking to a bus driver he`s known for some time. This driver was digging his garden last year [1996] when he hit something "very hard" with his spade. He dug it up to find that it was a "beetle" with an undamaged carapace; it was the size of a man`s fist and he found others. This year [1997] he found more but "a little smaller". I personally, can`t think of any native beetle the size of a man`s fist that can be hit very hard with a spade and go undamaged. The chap is being asked to draw a sketch, get a photo or an actual specimen. Any ideas? [Methinks, now, nodules of iron ore? - R] [4]

About a week later I got another letter from Mr Hooper, which said:

> 'I will, however, pass on any details [of the insects - R] I get to you so don`t worry you won`t miss out! I guess with the SE having a drier climate these days ("dust bowl of England" and all that!) exotics are bound to creep in more. The fact that the beetles of `96 were the size of a man`s clenched fist but this year there are smaller (but still large) one`s indicates a breeding population of something!'

[Mr Hooper, if you are reading this, I am not trying to be nasty; I just have a strange sense of humour - a breeding population of iron nodules?! Mind you, I had a neighbour in Wiltshire who believed stones grew. R] [5]

Lastly, in my penultimate blog on this subject, a visit to Hong Kong: on the CFZ website, *Mystery Insects* by Jon Downes, which I featured several days ago, mention is made of the "blood sucker":

> '...I have another mystery, which perhaps some reader of this article could help me solve. I was a child in Hong Kong during the nineteen sixties and I collected and kept a lot of local invertebrate wildlife. There was a small arthropod, (I think it was an insect, but after thirty years I cannot be sure), which the local children called the "blood sucker".
[including myself - R]

'It was about an inch and a half in length and appeared like a very thick set ant with `knobbly` legs and a fairly heavy chitinous covering. I suspect that it may have been the immature form of one of the small ground mantids [this is the conclusion I reached – R] but I am not sure. If anyone reading this either lived, or lives in Hong Kong, and could help me solve a problem which has been `bugging me` (if you`ll excuse the dreadful pun) for many years I would be extremely grateful.' [6]

[EDITOR'S NOTE: It turned out to be *Tricondyla pulchripes*, a specialised and flightless species of tiger beetle]

1. R. F. M. White Earwig. *Science Gossip* September 1st 1865
2. R. Highfield. Entomologist finds Britain`s 640th spider January 20th 1995
3. Anon. Spiders on speed get weaving. *New Scientist* April 29th 1995
4-5. Letters from Terry Hooper to Richard Muirhead August 6th and 12th 1997
6. Jonathan Downes. Mystery Insects. CFZ Website Accessed April 15th 1998.
                        Since updated.

Peter Gabriel – `Biko`.

September `77, Port Elizabeth, weather fine
it was business as usual
In Police Room 619
Oh Biko, Biko, because Biko
Yihla Moja, Yihla Moja
the man is dead, the man is dead.

MONDAY, NOVEMBER 16, 2009

MUIRHEAD'S MYSTERIES: Lake and sea monster archives part one

Fellow cryptozoologists and Forteans,

Today I present a series of notes from my archives (which roughly, in my system of filing, includes items up to and including the year 1950) on sea and lake monsters and strange creatures.

This first item is from *The Naturalist's Note Book* for 1867:

> "A CURIOUS FISH.-The Courier de Saigon brings, as a contribution to Natural History, the not very credible-sounding description of a fish called "Caoug" in the Anamite tongue, which is said to have saved the lives already of several Anamtes;for which reason the King of Anam has invested it with the name of "Nam hai dui bnong gnan" (Great General of the South Sea).This fish is said to swim round ships near the coast,and,when it sees a man in the water, to seize him with his mouth and to carry him ashore. A skeleton of this singular inhabitant of the deep is to be seen at Wung-tau,near Cape St.James.It is reported to be 35feet in length, to have tusks "almost like an elephant," very large eyes, a black and smooth skin, a tail like a lobster, and two "wings" on its back. [1]

I'd like to track this cryptid down if at all possible.

The *Irish Builder* of May 15th 1890 had the following piece:

> "IRISH LOCAL LEGENDS. No XV.-LEGEND OF THE WATER SERPENT OF LOUGH DERG, COUNTY OF DONEGAL.
>
> In the old pagan times a peistha, or water serpent, of immense girth and of still greater trail, was believed to haunt a celebrated spot in the northern part of Ireland. When St Patrick landed in Saints Island on Lough Derg, this large water serpent was known to have tenanted its waters. But the saint could not tolerate the presence of such a monster, and accordingly with a stroke of his staff the peistha was destroyed [2]

Now, the cat-headed sea-serpent, from *The Country-Side* of June 15th 1907. Unfortunately some of the words on the photocopy I made are indistinct:

> "Cat-headed sea serpent.- The officers of? Canarder Campania, on the arrival of the ? at Liverpool, on May 25th, gave a vivid description of a sea serpent which was seen Friday morning, when the ship was off the ? Coast.The creature, which rose to the? only a hundred feet from the ship's ? was roughly sketched by one of the officers. Eight feet of the front part of the ? which was the shape of a python, stood ? out of the water, surmounted by a head resembling that of a cat. The tail projected ? feet out of the water, and there was a space about thirty feet between head and tail, so ? the officers compute the length of the animal at about forty feet. H.P.G." [3]

Finally, a story from Canada:

> "Dishonest Sea Serpent. Residents of Amherst Island, Ontario, are alarmed by the persistent rumours that a monster sea serpent is in the vicinity. People who spend much of their time along the shore or on the water declare that the serpent makes regular appearences, that it is unlike anything ever seen before in this locality and is of a pugnacious nature. The result of these stories is that few dare to venture out without some sort of weapon to be used in case of attack. The wharves are becoming deserted; and

whether the serpent can justly be blamed or not, it is certain that foodstuffs left near the water's edge have disappeared wholesale. Large milk cans on several occasions have been carried off,and while no one has actually seen the monster take them, it is commonly believed that he is the thief." [4]

1. *The Naturalist's Note Book* 1867 p.167
2. Anon.*The Irish Builder*. May 15th 1890.p.115
3. H.P.G.Cat headed Sea Serpent.*The Country-Side* June 15th 1907
4. Anon.Dishonest Sea Serpent. *South China Morning Post*.September 26th 1912

Richard belongs to Jayne
And Jayne belongs to yesterday
How can I go on
When every alpha particle hides a neon nucleus

Neil belongs to love
And love belongs to no man
How can he go on
When no-one answers the adverts in his mind? – `Richard` by *Billy Bragg*

TUESDAY, NOVEMBER 17, 2009

MUIRHEAD'S MYSTERIES: Lake monster archives part two

Hi folks,

Here is the second and final part of lake monster archives, these items were illustrated in their original publications. Firstly, an item from *The Illustrated London News* titled `Another Sea Serpent`, which was actually in the form of a letter from several passengers aboard the *Imogen*, in the [English ?] Channel on April 15th 1856:

'Sir, We beg to hand you the enclosed sketch of a Sea-Serpent we had the good fortune to sight on the 30th March last. Imogen, from Algoa Bay, towards London. Sunday 30th March, 1856. Lat 29 deg. 11 min N. Long 34 deg.26 min. W. bar 30.50 calm and clear. Four vessels visible to southward and westward. About five minutes past eleven, am the helmsman drew our attention to something moving through the water, and causing a strong ripple about 400 yards distant from our starboard quarter. In a few moments it became more distinct...and showing an apparent length of about forty feet (above the surface of the sea), the undulations of the water extending on each side to a considerable distance in its wake. Mr Statham immediately ascended to the maintop sail-yard, Capt. Guy and Mr Harries watching the animal from the deck with the telescope. After passing the ship about half-a-mile, the serpent "rounded to" and raised its head, seemingly to look at us...then steered away to the northward...possibly to the neighbourhood of the Western Islands, frequently lifting its head...We traced its course

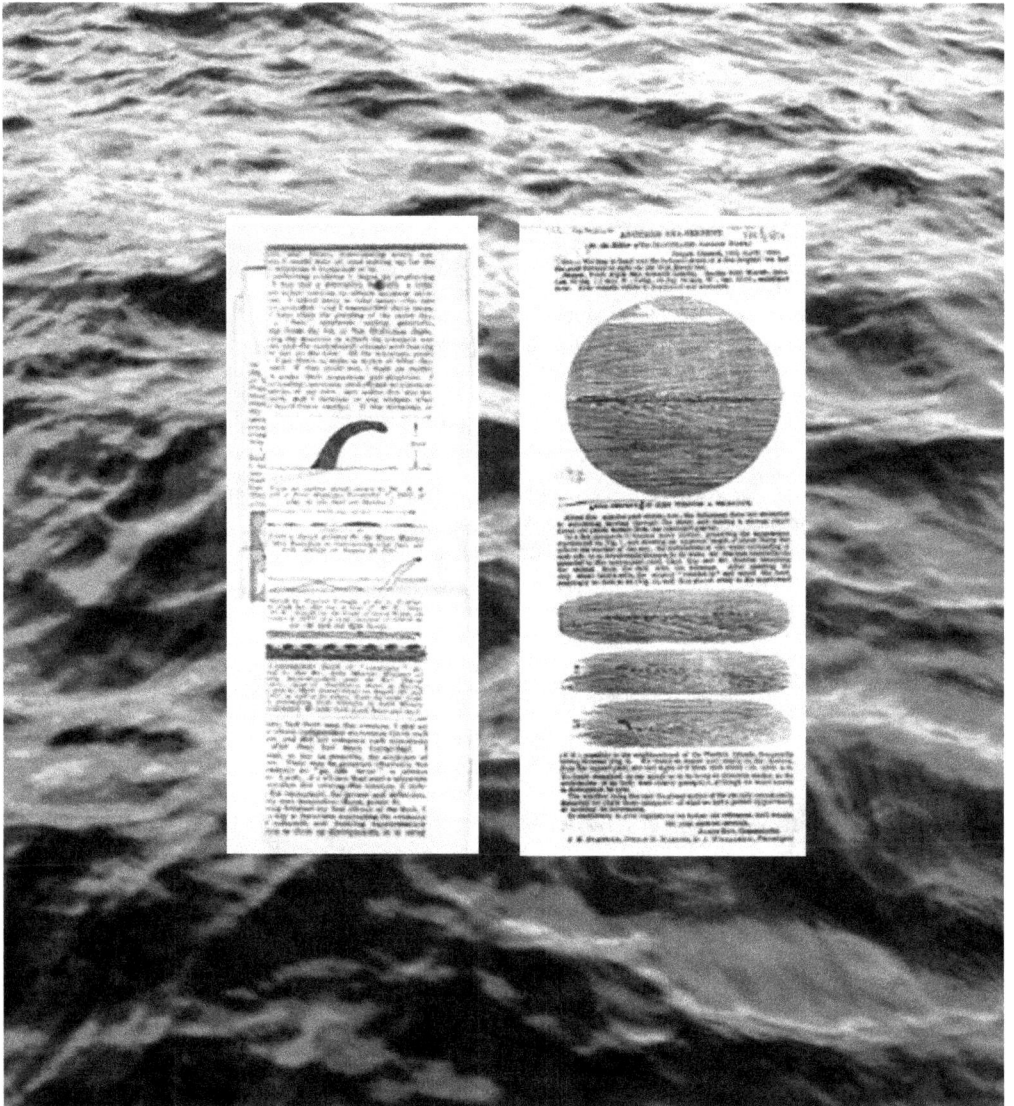

until nearly on the horizon, from the topsail-yard, and lost sight of it from deck about 11h.45am.`

No doubt remained on our minds as to its being an immense snake, as the undulations of its body were clearly perceptible, although we were unable to distinguish its eyes. The weather being fine and the glassy surface of the sea only occasionally disturbed by slight flaws (cats paws) of wind we had a perfect opportunity of noticing its movements. In conformity to your regulations we inclose our references, and remain Sir, your obedient servants, James Guy, Commander, J. H. Statham, Julian B. Harries, D. J. Williamson, Passengers. [1]

By December 1933 the 'modern' phase of the Loch Ness Monster scare was well underway. By 'modern' I mean as opposed to famous pre-twentieth century sightings such as by St Columba in 565AD as recorded by his biographer, Adamnan. The Scottish *Daily Express* reported on June 9th 1933:

'Mystery fish in Scottish Loch - Monster reported at Fort Augustus. A monster fish which for years has been somewhat of a mystery in Loch Ness was reported to have been seen yesterday at Fort Augustus.' [2]

June 28th 1933:

'Two men and a woman who were boating on Loch Ness had an unpleasant and exciting experience today. The `monster` rose out of the water about 50 yards from where the boat was drifting. One of the women fainted.' [3]

August 12th 1933:

'An effort to photograph the Loch Ness Monster is to be made by Captain Ellisford, a well known amateur photographer. He arrived at Inverness today with a large box of modern photographic material. He will use a telephoto lens.' [4]

By December 9th 1933 *The Times* had the headlines 'The Loch Ness Monster – A Survey of The Evidence – Fifty-One Witnesses. By Lieut-Commander R. T. Gould.' Gould speculates as to what the creature in Loch Ness could be and how it got into the loch. He describes his survey of the loch and his methodology with regards to interviewing eye-witnesses. Gould also speculates on what the Loch Ness Monster could be with headlines in the article such as A "Prehistoric" Neck and A Huge Marine Newt? [5]

1.   J.Guy et al *Illustrated London News*. May 3rd 1856
2-4. N.Witchell. *The Loch Ness Story*. (1982) p.40
5.   R.T.Gould. *The Times* December 9th 1933

SATURDAY, NOVEMBER 21, 2009

MUIRHEAD'S MYSTERIES: A Fortean at a friary

Dear folks,

Muirhead`s Mysteries is back after my five day spell at the Friary of Saint Francis at Alnmouth on the coast of north-east England. Those of you living abroad may have heard of the record-breaking rainfall we have had here. Alnmouth missed the worst of the rain and wind, which affected north-west England and Ireland.

As far as Forteana and cryptozoology are concerned, it was a fruitful occasion. As far as spiritual matters go, unfortunately I was too worried about various things to make much progress, but that's another story. The things I was worried about had nothing to do with the CFZ I hasten to add.

In chronological order, I start with an undated item. Near the entrance to the Friary there is a framed print comprised of a map of north-east England with small drawings and handwritten comments about these drawings. This item is on that framed print – 'At Whittle Quarry a toad was found in a stone where it had lived 1,000,000 years (they say). It immediately died.' [1] Unfortunately the date when it was found and who 'they' were are not mentioned.

For the next item, the date is given:

"1544- The famous Bamburgh sea-devil. [2] "

Obviously it must have been famous at one point.

"In 1765 a woodcock carrying a great diamond was shot." [3] This happened near "Mary's Isle"

Now jumping ahead to the 20th century; I met a man from Hereford who told me that he had heard that about 10 to 20 years ago a member of the SAS had seen a single wild cat (as opposed to a puma or panther) in the Brecon Beacons area of Wales. The wild cat was supposed to have died out in Wales by 1870! I have left the man from Hereford my email address and he is going to contact me if he finds out any more information.

The *Church Times* of November 13th 2009 had a charming but sad story of the death of an albino squirrel in Dorking, (Surrey?). The story goes:

'Conspicuous absence: flowers have been left in the churchyard of St Martin's Dorking, in memory of an albino squirrel that lived there until it was run over by a van last week. The squirrel, known to locals as Albi, Snowy or Percy was a popular figure in the town centre, and "peacefully co-existed" with the grey squirrels for about three years, said the Vicar of St Martin's the Revd Richard Cattley. "Children used to stop and watch him; he was quite a character. He got quite tame, and people would get quite close to him."' [4]

Lastly, a strange bird story:

'Budgies mystery. Bird lovers and exhibitors watched on in horror as 38 prize budgies keeled over and died during a show Fanciers feared a gas leak caused the tragedy but after investigations by experts there was no explanation for what happened at the village hall event in Gwynedd, North Wales.' [5]

1.   Framed picture. Friary of St.Francis. Alnmouth. November 2009.

2.   Ditto.
3.   Ditto.
4.   Untitled story about albino squirrel. *Church Times*. November 13th 2009.
5.   Budgies mystery. *The Journal*. November 17th 2009.

And now to finish off with, an appropriate selection of lyrics from `Flood 2` by The *Sisters of Mercy*

> And her hallway moves,
> Like the ocean moves,
> And her hallway moves
> Like the sea
> Like the sea
> She says no no no no harm will come your way
> She says bring it on down, bring on the wave
> She says nobody done no harm
> Grace of God and raise your arms
> She says face it: and it's a place to stay..

SUNDAY, NOVEMBER 22, 2009

MUIRHEAD'S MYSTERIES: Lake and sea monster archives part three

Hiya folks,

Today I present more archival information relating to sea and lake monsters, with some information from Ireland. I just picked out a small batch of photocopied information from my Lake and Sea Monsters file and immediately found an interesting link to another collection of material I had passed on to Jon just before his famous trip to County Kerry in Ireland earlier this year.

I do remember that this information mentioned roaring eels. Less familiar was the mention of a black man' diving for a 'carbuncle' in a lake in Co. Kerry. Now, this carbuncle was not definition 1.

'1. An extensive skin eruption, similar to a boil, with several openings
'2. A rounded gemstone, esp. a garnet cut without facets.' [1]

It was alleged to be a kind of Lake Monster with similarities to definition 2. above. I quote:

'There is another jewel in the Kerry crown, an animal unique, known only there, a resplendent creature called the Carrabuncle. It is true that it has never been seen by the cold critical eye of science, but Matthew Arnold reminds us through the mouth of Empedocles that "much may still exist that is not yet believed." The Carrabuncle is mentioned in Charles Smith`s *Antient and Present State of the County of Kerry*, published in 1756, as having been seen in the Killarney Lakes - but he erroneously assumed that the name he heard belonged not to an animal but to the familiar precious stone. Henry Hart, when exploring the Kerry mountains in 1883, came on its track again, this time on Brandon. He learned that its home was in Lough Veagh (Loch Betha, birch lake) where the people gather fresh-water mussels for the pearls which some of them contain [and which also exist in the Cladagh River in N. Ireland]. "These come off an enormous animal called The Carrabuncle, which is often seen glimmering like silver in the water at night. This animal has gold and jewels and precious stones hanging to it, and shells galore; the inside of the shells shines with gold." Five years later, Nathaniel Colgan happened on Hart`s informant, when climbing on Brandon, and obtained further particulars [2].

The true home of the Carrabuncle, it appears, is Lough Geal (Loch geal, shining lake), not Loch Veagh. His informant had never seen it, but if you could only catch it you would get some things of great value that follow after it...it is seen, it would appear, only once in seven years, and then it lights up the whole lake...But a very interesting side issue to this legend, as yet unexplained, pointed out by Colgan, arises from the reference by Alfred Russel Wallace in his *Travels on the Amazon and Rio Negro*, to the Carbunculo, a mythical animal of the Upper Amazon and Peru.Colgan was puzzled as to the connection (if any) between the Irish and the South American creature. The probable explanation of the presence of the name in these two widely separated areas lies in the fact that both regions were in intimate connection with Spain...It would seem that Carrabuncle corresponds similarly to the Spanish Carbunculo, meaning the precious stone we call carbuncle - though how the word came to be applied either in Ireland or on the Amazon to a water-monster is not clear.' [3]

So what have we here? A rare or cryptid oyster/mussel? Or did it get to Ireland via Spain or Brazil? Bear in mind the first Irish came to Eire via the Mediterranean, or so I`ve heard. This book was first published in 1937 so records of it now should not be impossible to trace. Could the black man have had some connection with the Amazon?

The next story is also from Kerry, but on the coast near Dingle:

'Strange legends hang around this wild and primitive coast connected with unknown monsters of the deep. "It is my belief," said a man one day, "and the belief of many, that there is no animal on land but what has its like in the sea"; and then he proceeds to tell of a strange creature which goes around the coast-line from Magharees to Brandon Head, and is called by the people the "Currane Duv," or "Black Sow." It has been seen in the memory of man - a large animal, 15 feet long, "with mane like a horse", a foot in length, which waves in the water as it swims. Sometimes it goes up river for a short distance, but its chief habitat is the sea, where it is a terror to the fisher folk! [4]

So there we are! Any opinions from Irish cryptozoologists? Ronan?

1. Collins Dictionary and Thesaurus. (2004) p.171
2. See *Irish Naturalist*.vol.23 p.59 1914.
3. R.L. Praeger. *The Way That I Went*.(1937,1939,1947,1969,etc)pp 364-366.
4. C.P. Crane. Kerry. (publication date unknown.)p.227.

As weather conditions here are still very poor, I conclude with `The Storm` by *Big Country*.

I came from the hills with a tear in my eye
The winter closed in and the crows filled the sky
The houses were burning with flames gold and red
The people were running with eyes filled with dread
Ah, my James
They didn`t have to do this
We chased them for miles I had hate in my eyes
Through forest and moors as the clouds filled the skies
The storm broke upon us with fury and flame
Both hunters and hunted washed out in the rain

1 COMMENT:

Ego Ronanus said...
Currane Duv would be an anglicised form of crain dubh, which in Irish literally means"black sow". The creature you describe is rather more like Irish lake monsters than sea monsters.

TUESDAY, NOVEMBER 24, 2009

MUIRHEAD'S MYSTERIES: Steller's Sea Cow

Hi again, fellow cryptozoologists and Fortean Zoologists, today`s blog should really be subtitled the Steller`s Sea Cow and Sea Monkey blog because it is these subjects I will be examining

Steller`s Sea Cow has been defined as:

"A large Sirenian of the North Pacific Ocean, presumed extinct since 1768. Scientific name: (*Hydrodamalis gigas*), given by Eberhard Zimmerman in 1780.

Variant name: Kapustnik (Russian ,"cabbage eater")
Physical description: Length, 20-26 feet,
Weight: up to 10 metric tonnes.

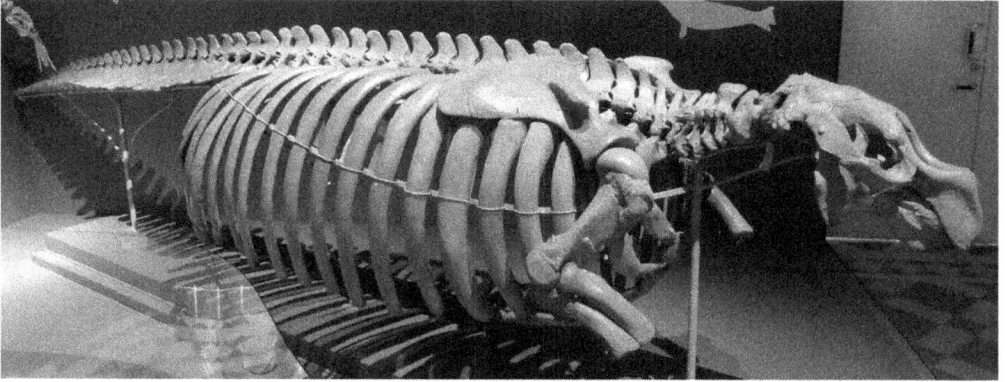

Description: Tough, dark brown skin. Rotund body. Small head. No functional teeth. Bilobate tail.

Behaviour: Average submergence time, four to five minutes. Strictly a seaweed-eater. Distribution: Gulf of Anadyr, Siberia; Commander Islands in the Bering Sea; Attu, Alaska. Significant sightings: A E Nordenskiold interviewed several residents of Bering Island who affirmed that sea cows were still being killed in the late 1770s. Around 1854 two other natives, Merchenin and Stepnoff, apparently saw an animal in the ocean that spouted water from its mouth".

Polish naturalist Benedykt Dybowski was certain that sea cows had survived off Bering Island as late as 1830. Lucien Turner interviewed an Aleut woman who said that her father had seen sea cows off Attu in the Aleutian Islands, Alaska, in the mid-nineteenth century. A sea cow allegedly was stranded on the shore of the Gulf of Anadyr, Siberia, in 1910. In the early 1950s, a harpooner named Ivan Skripkin told of 32-foot, finless animals that appeared every July not far from Bering Island. The crew of the Russian whaler *Buran* observed six dark-skinned marine animals, 20-26 feet long, feeding in a lagoon near Cape Navarin, Chukot Autonomous Province, Siberia in July 1962. They had small heads, bilobate tails, and bifurcated upper lips.

Russian fisherman Ivan Nikiforovich Chechulin walked up to and touched a live sea cow in the summer of 1976 at Anapkinskaya Bay, south of Cape Navarin. Its tail was forked like a whale's, and it had a long snout". [1]

So the above is like, say, the skeleton or frame-work of the sea cow. The human element is no less interesting: In 1741 two ships, the St Peter and St Paul were travelling past the Aleutian Islands. On one ship the Danish explorer Vitus Bering "was wrecked on Attu, the westernmost island in the chain. Those members of his crew who didn't die of exposure managed to stay alive by harpooning and eating vast sea animals – up to 30 feet long – which were feeding in the kelp around the islands. These huge mammals, members of the Sirenia family – which today includes the manatees and the dugongs, all rare or endangered – became known as Steller's sea cows. Over the past 20 years there have been various reported sightings of the sea cow off the western Aleutians and Kamchatka. The descriptions are intriguing, but it is very possible that the animals were whales, or walruses, which only seldom venture as far south as the Aleutians. [2]

Furthermore,

> "They [the Russians - R] killed the slow moving sea cows in such numbers that there were none left by 1768. (We have no way of knowing how many sea cows were on the islands [the Commander Islands - R] when Bering landed, but Leonhard Stejneger, Steller's biographer, has estimated that there were some fifteen hundred.) It had taken only twenty-seven years for the Russian adventurers to eliminate the hapless sea cow from the face of the earth, but the sealers had no way of knowing that this was the last of them; they probably assumed that there were similar undiscovered islands with more sea cows". [3]

Sven Waxell, an officer on Bering's 1741 voyage aboard the St Peter which was wrecked on the Commander Islands drew the only known example of a sea cow. [4]

Now the Sea Monkey: [not to be confused with the strange semi-animate bogus beings made of chemicals of this name to be bought in Hong Kong in the 1970s - R] According to Steller

> "During this time we were near land or surrounded by it we saw large numbers of hair seals, sea otters, fur seals, sea lions, and porpoises...On August 10 [1741] we saw a very unusual and unknown sea animal, of which I am going to give a brief account since I observed it for two whole hours. It was about two Russian ells [six feet] in length; the head was like a dog's, with pointed erect ears. From the upper and lower lips on both sides whiskers hung down which made it look like a Chinaman. The eyes were large; the body was longish round and thick, tapering gradually towards the tail. The skin seemed thickly covered with hair, of a grey colour on the back, but reddish white on the belly; in the water, however, the whole animal appeared entirely reddish and cow-coloured. The tail was divided into two fins, of which the upper, as in the case of sharks, was twice as large as the lower. [5]

> "No record of any other sighting of this animal has been found, and so scientists have generally agreed that it must have become extinct at around the same time as the sea cow. That was until 1965...On a clear afternoon in June, [when Brigadier Smeeton, his wife and daughter and Henry Combe - R] were sailing about four miles off the northern coast of Atka, in the central Aleutians, bound for Deep Cove, when they saw a strange animal which they couldn't identify". [6] Its appearance was like the Sea Monkey. The Sea Monkey has variously been identified as a otter or a wayward Hawaian monk seal.

Well that's all folks. See you tomorrow.

1.	G.M. Eberhart *Mysterious Creatures A Guide To Cryptozoology vol. 2* N-Z.(2002) p.519
2.	M. Clark Salty Tales. *BBC Wildlife Magazine*. January 1987. p.11
3.	R. Ellis *Monsters of the Sea* (1995) p.p93-94
4.	R. Ellis. Ibid.p.94
5.	G.W. Steller *Journal of a Voyage with Bering*, 1741-1742. (1988 ed) in R.Ellis Ibid p.97
6.	M. Clark op cit p.12

Devo – 'Patterns'

Patterns all around you
Patterns everywhere
Patterns of behaviour
Sometimes seem unfair
Can you recognise the patterns that you find?

1 COMMENT:

Retrieverman said...
In Farley Mowat's Sea of Slaughter, page 171, there are some accounts of sea cows that appeared in the North Atlantic coast of North America.

Richard Whitbourne to have come across one in St. John's harbour (Newfoundland). John Josselyn described coming across one in Casco Bay (Maine) in 1670.

These animals supposedly were known to have occurred off the coast of Greenland. Now, the first two could have been errant West Indian Manatees, because there is no description of huge size accompanying them.

WEDNESDAY, NOVEMBER 25, 2009

MUIRHEAD'S MYSTERIES: Lake and sea monsters part four

Hi again.

Today in my final look at lake and sea monsters I'm going to travel to Australia. I have recently come across a collection of material on Aussie 'Loch Ness-type monsters and a sighting or rather viewing of a possible 'monster' in Loch Alexander within Darwin city council's remit. This happened in mid-September 2009. Of course there is the Bunyip, but I will not be looking at it in today's blog. For those interested in it there is good information on it in Healy and Cropper's book *Out of The Shadows* [1]

Rex Gilroy is Australia's authority on the Hawkesbury river monster(s) near Sydney: "Loch Ness-type monsters are alive and swimming in the Hawkesbury River near Sydney. Controversial Catacomb naturalist Rex Gilroy has more than a score of carefully documented eyewitness accounts of strange reptile-like creatures seen in the river over the past 20 years. I have absolutely no doubt, after piecing together the evidence from all these independent sources, that what people actually saw was the plesiosaurus…"

"Rex Gilroy also claims the monsters are alive and well in Lake Taupo, near Whaiteko, in New Zealand's North Island. On a recent trip there, with his wife Heather, Gilroy photographed what he believes is a plesiosaurus…Both my wife Heather and myself saw a long dark brownish shape a few hundred metres from the shore where we standing, moving across the lake…At least 15 metres of body length could be detected but there was no body outline…Rex Gilroy has investigated several Maori legends about water monsters and says they tally with the scientific description of the plesiosaurus or 'Nessie'. And there are also Aborignal legends about the monsters of the Hawkesbury, he says. But not all the 'Nessie' long necks seen in the Hawkesbury River are gigantic creatures. " Take the case of 'John' who was fishing near Ettalong in March 1971 with his father. He caught a strange little beast, about 45 centimetres long with a long paddle-like tail and fins. It had a snake-like head, and opened its mouth and hissed when 'John' and his father attempted to free it.

"They kept the strange beast for some days, hoping someone could identify it. No one did. It was only later when they saw pictures of a plesiosaurus in a book that they realised they might have thrown a baby one away- and so missed their chance of making world history." [This is interesting. Isn't there a story of a "baby" Cadbosaurus ?]…Gilroy's efforts have earned him a place in Von Daniken's books where his claims are accepted. He would like to hear from anyone who has seen anything mysterious in the Hawkesbury River". [2]

The magazine then gives Gilroy's address, which I reproduce here, but please bear in mind this is dated 1981:

REX GILROY
KEDUMBA NATURE DISPLAY
DEDUMBA EMPORIUM
ECHO POINT ROAD
KATOOMBA
NSW 2780

"Gilroy will reappear in a minute but we need to take a quick look at the image caught on Google Earth by security guard Jason Clarke." The images captured on Google Earth show a large brown object in the murky water at Lake Alexander, where a giant cod bit a woman on the foot last month [can I have cod, toe and chips please,?? The toe's salty enough ho ho - R] …Security guard Jason Cooke said the 65ft oblong shape followed by thin strands is actually an image of the possibly mythical creature supposed to inhabit Loch Ness in the Scottish Highlands…We estimate our own "mystery" Loch object or beast to be

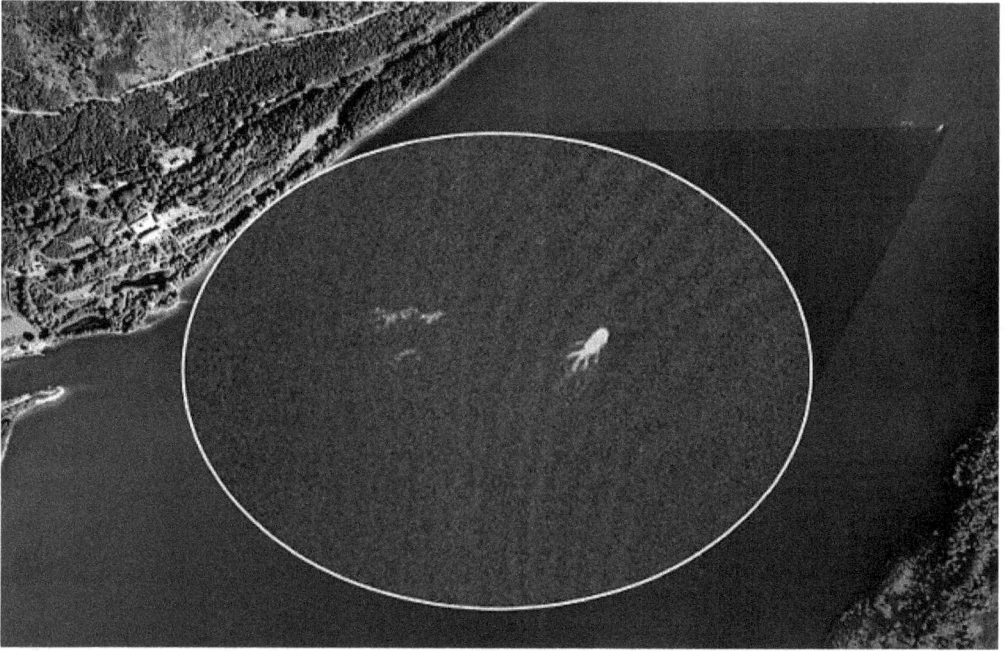

approximately 5m long or wide…Two weeks ago Sydney cryptozoologist Rex Gilroy said he had sighted Sydney's own "Nessie", claiming he witnessed a 12m giant surface in the Hawkesbury River. Through binoculars Mr Gilroy saw a dark shadow "with a longish neck" near Wiseman's Ferry…After hearing of the Hawkesbury Monster in 1965 he found accounts dating back to pre-colonial times, with stories told of children being attacked by the "moolyewonk" When fishing boats were found overturned and the occupants missing in the 1980s, the Hawkesbury Monster was the prime suspect."[3]

That's all, folks. See you tomorrow. Richardo.

1. T. Healy and P. Cropper. *Out of The Shadows. Mystery Animals of Australia.* (1994) Bunyip. Chapter 6 is pp159-180
2. G. Lyons. `Nessie Down Under`. *People.* June 17th 1981. pp3-5
3. M. Cunningham. *Satellite shots reveal a mysterious lake monster.* http:// www.ntnews.com/ [accessed November 24th 2009]

The Jam `Smithers Jones`. Their classic indictment of capitalism

Here we go again, it's Monday at last,
He's heading for the Waterloo line,
To catch the 8am fast, its usually dead on time,
Hope it isn't late, got to be there by nine.
Pin stripe suit, clean shirt and tie,
Stops off at the corner shop, to buy The Times

`Good Morning Smithers-Jones`
`How's the wife and home?`
`Did you get the car you've been looking for?`...

2 COMMENTS:

Dale Drinnon said...

Not only is there rock art in Australia that exactly reproduces Plesiosaurian anatomy (and traditional reproductions of the same) but you get the same thing in New Zealand. This includes the precise details of the skeletal structure of the flippers, limb girdles and abdominal ribs, represented "X-ray" style, as well as representations of the living creatures intact.

That being said, the most common "Bunyips" in both areas are ordinary seals, but they range up to Elephant seals in size.

Dale Drinnon said...

Subsequent to the comment I posted, Lindsay posted a blog about the Australian Aboriginal depictions of Plesiosaurs and I followed with a posting on the similar pre-contact Maori rock-art depictions from New Zealand.

WEDNESDAY, NOVEMBER 25, 2009

MUIRHEAD'S MYSTERIES: Dragons

In the depths of my archives I found a story titled `The Dragon Farm` from a book called *Avery Memorial*. It was written by someone calling themselves 'C.C.C.' All other details are unknown. There is an Avery near Midhurst in West Sussex and one near Southampton. I have a vague memory of finding this short account whilst working in Warwickshire in the mid 1990s but I am not absolutely sure. However, the mention of Alcester (`Roman fort on the River Alne`) would seem to confirm the Warwickshire connection. The following is an abridged version.

'Though I had often heard of The Dragon Farm and the curious chimney-piece preserved there, it was not until recently that I had an opportunity of paying it a visit. The house I found to be an old half-timbered one, standing upon a base of stone-work, and moated on its hill side. From the number of perpendiculars in the timber work and the form of the angle-pieces it would appear to be a place of some antiquity, though not belonging to the Saurian period as the local tradition would imply.

'The place derives its name from two so-called dragons, carved in oak, on an otherwise plain chimney-piece in one of the rooms, and the story told of them is as follows: - In that extinct age when the dragon and the wyvern, the cockatrice and the fire-drake, to say nothing of gigantic serpents, griffons &c, existed upon the earth, there lived at the farm house a certain man who had large flocks of sheep. To his great mortification, however, these, instead of increasing, decreased in number to an alarming extent. Whither the missing ones went he could not tell, and the rangers, woodwards and verdurers (?) of Feckenham assured him that they had not strayed in the forest; neither could he find about his own pastures any trace of their having been torn by wolves. In his distress he went to consult a holy man who simply bade him "Watch and pray", and he therefore set his shepherds to watch while he betook himself to his beads. On the night following, the shepherd came to his master and told him that two evil beasts were dragging away some lambs, whereupon the men armed themselves with (with "guns" says one of my informants) and were quickly in pursuit. They followed the ravening creatures and saw them disappear with their prey through a hole in the butt of an enormous oak tree. Having found the den of the spoilers they quickly shot them, and the flocks thenceforth began to increase. On examining the tree they found the trunk quite hollow, and there was room within to turn a coach round!

'It was in commemoration of this strange occurrence, says the tradition, that the two so-called dragons were carved over the chimney-piece at the farm. An inspection, however, of the carving will show any one versed in such matters that it is a piece of very good work of the time of James I. Did not the date 1614 on a shield inform us positively of this?

'The creatures themselves should have been described as serpents rather than dragons; their tails, barbed at the ends, are interlaced, and the eyes in their regardant heads glare fiercely at each other; the teeth, however are those of a crocodile, though the tongue ends in what is intended to represent a sting. This kind of nondescript is not unfrequent in our neighbourhood and examples may be seen at Tookey`s Farm, at Alcester, and elsewhere.' [1]

A web site about the Ropen, Papua New Guinea`s supposed living pterosaur, says: 'What do dragons and pterosaurs have in common? Celtic dragons had arrows at the end of their tails, which may relate to pterosaur tails. What about Rhamphorhynchoid pterosaur tails? Are not dragon tails also long? Perhaps most noteworthy are the wings: both pterosaurs and flying dragons have featherless wings' [2]

1. C.C.C. No. 247. *The Dragon Farm.* Avery Memorial. Date unknown
2. Are Dragons Pterosaurs? http://www.objectiveness.com/dragons-1/ [accessed November 25 2009]

Talking Heads `Animals`

I'm mad….And that`s a fact
I found out…Animals don`t help
Animals think…They`re pretty smart
Shit on the ground…See in the dark
They wander around like a crazy dog
Make a mistake in the parking lot
Always bumping into things
Always let you down down down down

FRIDAY, NOVEMBER 27, 2009

MUIRHEAD'S MYSTERIES: Early Chinese knowledge of the Kting Voar, a recently discovered bovid from Cambodia

The early 1990s was a good time for cryptozoologists and other scientists in south-east Asia, particularly in Cambodia and Vietnam where the Kting Voar, "also known as the Khting Vor, Linh Duong or Snake-eating Cow *(Pseudonovibos spiralis)* is a bovid mammal reputed to exist in Cambodia and Vietnam…The Kting Voar is normally described as a cow-like animal with peculiar twisting horns about 45 centimetres (20 inches) long. Anecdotal descriptions of the animal mention a spotted pelage. Folklore claims that it has a connection with snakes.

Kting Voar is the animal`s Cambodian name. This was erroneously translated in the West as `jungle sheep`, leading to a mistaken assumption that the animal was related to sheep and goats. In fact the name means `liana-horned gaur` (a gaur is a species of wild Asian cow).

Adding to the confusion, the Vietnamese name linh duong, meaning `antelope` or `gnu`, was once reported to refer to this animal. However, this is in fact a local name for the Mainland Serow.

Other Kampuchean names possibly include kting sipuoh (`snake-eating cattle`) and khting pos.

"For Western scientists, the first evidence supporting existence was a set of horns found by biologist Wolfgang Peter in a Ho Chi Minh City market. (W. P. Peter and A. Feller. Horns of an unknown bovid species from Vietnam (Mammalia: Ruminantia) Faun. Abh. MusTierkd. Dresden 19,247-253.)...All supposed Kting Voar specimens that were subject to DNA analysis to date have turned out to be artificially shaped cattle horns...The most likely explanation, given the DNA results and the unusual spotted fur (which is well known in domesticated, but unknown in wild cattle), seem to be that modern specimens at least are cattle horns shaped by a complicated technique in order to serve as anti-snake talismans...There is also an earlier report of British tiger-hunters in the first part of the 20th century, who observed Kting Voar and shot two as tiger bait... The existence of the Kting Voar is far more likely than that of other cryptids. IUCN Red List of threatened species lists it as endangered, stating "The existence and systematic position of *Pseudonovibos spiralis* is currently being debated. There are undoubtedly manufactured trophies ("fakes") in circulation, but the precautionary principle requires us to assume that the species did exist and may still exist." [1]

In the abstract to their paper `Chinese sources suggest early knowledge of the `unknown` ungulate *(Pseudonovibos spiralis)* from Vietnam and Cambodia`, Alastair A. Macdonald and Lixin N.Yang stated

`A survey of historical Chinese encyclopaedias, compilations and textbooks from the Ming and early Qing dynasties (14th to 18th centuries) was carried out for information that might fit an animal from Vietnam and Cambodia which is known only from its distinctive horns. These horns have a raised, rib-like pattern of rings round much of their length, and a backward curl of the horn`s tip. One illustrated text found in the San Cai Tu Hui, a compilation of knowledge by Wang Chi and his son Wang Si Yi (1607), seems to bear a close resemblance to the information which has recently been gathered during field trips in Cambodia and Vietnam`. The authors conclude that additional information on endemic animals in the region may be found in the writings of that part of the world....

Results. Illustrations and brief descriptions of goat-like animals were found in many of the books and manuscripts consulted. Most of them clearly referred to species present in northern China and Mongolia. However, one illustrated text found in the San Cai Tu Hui (Wang Chi & Wang Si Yi,1607) seemed to bear a closer resemblance to the information which has been gathered in Cambodia and Vietnam. [2]

I hope Jon and I will be able to use these old Chinese encyclopaedias for our future book *The Mystery Animals of Hong Kong,* which we hope to start writing in a few years.

1. Wikipedia. Kting Voar. http://en.wikipedia.org/wiki/Kting_Voar [accessed November 26th 2009]
2. A.A. Macdonald and L.N.Yang. Chinese sources suggest early knowledge of the 'unknown' ungulate *(Pseudonovibos spiralis)* from Vietnam and Cambodia Journal of Zoology (1997) 241 pp 523-524.

Muirhead's Mysteries will be taking a short break until next Tuesday due to a hectic schedule.

Thanks to Darren Naish who provided me with the document on early Chinese knowledge of the Cambodian-Vietnamese ungulate

Bob Dylan `I Dreamed I Saw Saint Augustine`

I dreamed I saw Saint Augustine
Alive as you or me
Tearing through these quarters
In the utmost misery
With a blanket underneath his arm
And a coat of solid gold
Searching for the very souls
Whom already had been sold

TUESDAY, DECEMBER 01, 2009

MUIRHEAD'S MYSTERIES: A collection of Fortean zoological curiosities

Hello again,

Today I present a collection of Fortean zoological curiosities collected over the years from various British newspapers. There is no overall theme; they were really chosen for their strangeness. I hope you enjoy them as much as I did. Also, as you will see, the collection is bunched around the mid-1990s, the only reason being that this was the first batch I picked up from the murky, dark swamp of my files. And I'm supposed to be a qualified librarian! Likewise, there is no particular reason for the emphasis on caterpillars and butterflies; they just took my fancy, plus Jon and I have been working on a book about butterflies.

MICE SACRIFICE: 'Hundreds of thousands of mice are said to have drowned themselves in rivers in North-West China. Experts say the mass suicide could have been triggered by over-population, although others suggest the animals may have had a premonition of disaster such as earthquake.' [1]

KILLER CATERPILLAR: 'A venomous, hairy species of caterpillar whose sting causes burns and internal bleeding and can kill humans has claimed its fifth victim in three years in the southern Brazilian state of Rio Grande do Sul, a farmer's wife' [2]

A GEMMA OF A FIND: 'Nine-year-old Gemma Thorpe found a caterpillar of a kind last seen in England in 1949 in her grandfather's garden in William Road, St Leonard's East Sussex. Moth expert Colin Pratt from Hastings Museum identified the 3 ½ in black, red and cream caterpillar as a spurge hawk moth. He said: It's a freak accident. This caterpillar is normally only found on the Continent or in North Africa.' [3]

ARTY BIRDS: (This will make you smile.) 'Japanese psychologists have taught pigeons to discriminate between cubist paintings by Picasso and impressionist works by Monet, but they cannot tell a Cézanne from a Renoir.' [4]

EXTINCT BUTTERFLIES FOUND IN WOODS: 'Three butterflies - the pearl bordered and small pearl-bordered fritillaries and the wood white - which conservationists had believed were almost extinct in Britain have been discovered thriving in a secluded 20 acre wood near Godalming in Surrey' [5]

A BAD DAY FOR: 'Rats, hundreds of which were exterminated at the house of an elderly man in Florida who had kept them as pets. "They were all sizes," said a public health official, "from grandpa and grandma rats to young babies. These rats were healthy, obviously well fed." Their owner, Mr Angelo Russo, 76, has been taken to hospital for psychiatric evaluation.' (6) (That's a bit extreme, I hope I'm not taken to the local psychiatric hospital because of my large key-ring collection. Unless they're man eating key rings...HELP, WHAT'S THAT NOISE COMING UP THE STAIRS, ARGGGGH!!)

Er, where was I?

SMALL AND HAIRY: 'Researchers in Sumatra are looking for a 70 inch tall hairy dwarf, or a possible colony of dwarfs, in the Kerinici Slebat national park. "It's quite difficult to catch the dwarf," said a local, "because it always turns up alone and can run very fast."' [7]

A BAD DAY FOR: 'Namibian politicians who have been shouting at each other across the chamber of the National Council after the loudspeaker system had been rendered inoperative by rats chewing through underground cables.' [8]

Now for something slightly different: this is a quote from *New Scientist* on August 29th 2009, p.12 in *Creation Journal* of the Creation Science Movement vol 16 no.6 December 2009 p.8.

'Did two species mix to make butterflies? An egg that hatches into a caterpillar that then changes into a seemingly dormant chrysalis from which a butterfly emerges to lay eggs containing all the genetic information for the entire cycle is rather hard for an evolutionist to cope with. This is the result of an ancient hybridisation between an insect and a worm-like animal, according to zoologist Donald Williamson...Nobody knows where caterpillars came from, says Williamson, who thinks that many other invertebrate groups acquired their larvae in the same way. It is the only solution that makes sense.` "Tommyrot, other biologists snort. For a start, the resemblance between velvet worms and caterpillars is only superficial. "As appealing to the imagination as Williamson`s theory may be, it looks like the evidence is not there to support it." And they say that creationist ideas are an abuse of science! Tadpoles into frogs are observed but frogs into princes are not' [9]

That's nearly all, folks, except for one thing: I was talking to my Danish uncle Joergen the other day and I asked him if he had any books mentioning the Steller's sea cow, which I talked about in my recent blog. Well, Joergen didn't have any such books, but he revealed that Bering, upon one of whom's ships Steller was a passenger and after whom the Bering Straits was named, was Joergen's 10th great grandfather!

1   *Daily Mail.* August 12th 1993
2   *The Guardian.* January 13th 1994
3   *Daily Mail* August 31st 1994
4   *The Independent.* May 26th 1995
5   *The Independent* May 18th 1995
6   *The Independent* June 30th 1995
7   *The Independent* July 20th 1995
8   *The Independent* July 27th 1995
9   *New Scientist* August 29th 2009 in *Creation* vol 16 no 6 2009

Bob Dylan `The Ballad of Frankie Lee and Judas Priest`

Well, Frankie Lee and Judas Priest,
They were the best of friends.
So when Frankie Lee needed money one day,
Judas quickly pulled out a roll of tens
And placed them on a footstool
Just above the plotted plain
Sayin` "Take your pick, Frankie Boy,
My loss will be your gain."

WEDNESDAY, DECEMBER 02, 2009

MUIRHEAD'S MYSTERIES: A collection of Fortean zoological curiosities part two

Here we go again with a collection of more Fortean Zoological Curiosities from my archives. I hope you found yesterday's batch interesting. I will endeavour to make today's collection interesting as well.

CROCODILE IN TRAWL: Skipper George Milton and the crew of the Folkestone trawler *Number 9* when fishing were surprised by the weight of the trawl's contents. When brought on deck, the trawl was found to contain a dead crocodile, 12ft long. Cutting off the tail, which they took to Folkestone, the crew dumped the carcass into the sea. It is thought that the crocodile had died on a voyage in a steamer to England and had been thrown overboard. [1]

At least 6 months ago I tried to find out more about this without success.

MAN BITES DOG: 'An enraged Indonesian villager killed a dog by repeatedly sinking his teeth into its throat after it bit a young boy. Health authorities are checking the dog for rabies.'[2]

SUPER RATS ON MARCH IN ENGLAND: 'A hunt is on for an army of rats spotted on the march across southern England. The rats, 300 of them scurrying along in a column, were spotted by a night watchman at dawn yesterday near the Wiltshire town of Trowbridge. By the time a band of local authority rat catchers turned out, they had vanished. Local environmental officer, Mr Bill Grey, said: "It's very unusual for rats to move in such numbers." Naturalists have reported that a super-breed of rats, resistant to all normal poisons, is thriving in the rural counties of western England and starting to spread across the southern part of the country.' [3]

Swindon Advertiser

Pest control expert warns of invasion of super rodents

THE BOOM TOWN RATS

I wonder what became of these super-rats. Trowbridge is a rather Fortean town. I believe there was a shower of frogs there, c.1930s.

[EDITOR'S NOTE: The Wiltshire super rats resurfaced in 2016 as this rather wonderful front cover is testament]

DOG FRIGHT: 'A police dog is missing at East Hanney, Oxfordshire, after running from a screeching sound it heard in woods while undergoing night training.' [4]

I have written a note above this: November 20th 1964 Mystery animal seen at Nettlebed, Oxfordshire, Puma? *Alien Animals* [the book by the Bords]

DORMICE CLUE TO ROMAN BRITAIN: 'Discovery of the remains of three garden dormice not previously found in Britain has led archaeologists to believe that Roman Britain was not self-sufficient in grain. The presence of the dormice in the grain store discovered at South Shields suggests that they were taken there in bags of grain imported from abroad to feed the garrison.' [5]

GOOSE'S METEORIC FALL: 'A wild goose is thought to have been killed in mid-air by a falling meteorite. Farmer Duncan Tesloss, of Polebrooke, Northants, was watching a flock passing overhead. "Suddenly there was a blue flash," he says. "Something like a laser beam hit one of the geese and it dropped like a stone."' [6]

And finally…

FIRE FIGHTER: 'Firemen believe a blaze which damaged three homes in Paulsgrove, Hampshire, may have been caused by a starling carrying a smouldering cigarette end back to its rooftop nest.' [7]

1. *The Daily Mail.* May 26th 1926.
2. *The Guardian.* April 29th 1986
3. *South China Morning Post.* November 10th 1982
4. *The Guardian* October 25th 1986
5. *The Daily Telegraph* February 17th 1987
6. *The People* March 22nd 1987
7. *Daily Mirror* March 13th 1987

Sorry no time for song lyrics tonight, however I conclude with a poem I wrote at the age of 10. (Yes, I know its ageist, but give me a break, I was only 10!)

`OLD AGE`

Weak,poor,crippled,old
Eighty,empty,alone,
Biscuits,cat,silence,mould,
Dog,clock,groan.

3 COMMENTS:

Retrieverman said...
There is a problem with the theory on dormice, and I don't know why these experts didn't notice it. It is more likely that those dormice were brought with grain for another purpose.

They were actually brought to Britain to be eaten

The Romans loved to eat dormice and raised them for that purpose.

"One of the most intriguing of the Ancient Roman recipes is for the dormouse. Probably because the thought of it feels us with horror! The edible dormouse was farmed by the Romans in large pits or in terra cotta containers and eaten by the ancient Romans as a snack or as part of the first course of the Roman main meal called the Coena. Dormouse recipe serving instructions: Dormice were sprinkled with poppy-seed and honey and were served with hot sausages on a silver gridiron, underneath which were damson plums and pomegranate seeds."

http://www.roman-colosseum.info/roman-life/ancient-roman-recipes.htm

Retrieverman said...
Garden dormice (I didn't notice the word garden) are not the same species as the so-called edible dormouse.

Neither species is native to Britain, and the existence of edible dormice in the country happened only because some escaped from 2nd Baron Rothschild's estate at Tring (which is now a zoological museum of sorts.)

It is possible that the Romans could have brought garden dormice as a substitute species.

Oll Lewis said...

RE dog bites man, I was scanning in an article today for inclusion in the archives about a man biting a snake in Edinburg, Texas, USA. The fellow in question had been bitten by a coral snake so quick as a flash he grabbed the snake and bit it's head off and used the skin as a tourniquet to stop the venom spreading. Disgusting and cruel to animals to say the least, but it did save his life.

Anyway, I'm sure Alfred Harmsworth would approve, keep an eye out for that in a future archive update.

THURSDAY, DECEMBER 03, 2009

MUIRHEAD'S MYSTERIES: Fortean zoological curiosities part three

Hiya folks,

It's Muirhead's Mysteries time again.

Today we are on part three of Fortean shorts. I do a bit of bending the rules with item one because it concerns *Homo sapiens* not the usual 'animal' I have been covering, if you catch my meaning.

PEAK POPULATION: 'A remote mountain village where no-one has died since 1942 boasts 189 residents who are more than 130 years old - including one who is 142.' [1]

I love these tales of human longevity. Has anyone ever done a serious study of them? I plan to live to at least 200 myself.

OLD NEWS 'Workmen refurbishing a newsagent's in Halesworth, Suffolk, found a mummified black cat, buried 400 years ago to ward off evil spirits, under the floorboards.' [2]

There is a pamphlet called *Skulls, Cats and Witch Bottles*, by Nigel Pennick, that I gave to Jon years ago, which is well worth reading if you can find a copy.

RADIOACTIVE BATS: 'Bats contaminated by radiation from a nuclear dumping ground have invaded a children's holiday camp in Chelyabinsk, Siberia, the ITAR-TASS news agency said yesterday.' [3] I wonder how many more bats have been spreading radiation around other former Soviet nuclear dumps?

BODY OF BEHEMOTH: 'A journalist in northern Russia said yesterday he had discovered the remains of a huge "forest monster" which climbed trees and lived off bark, Itar-Tass news agency reported. Vyacheslav Oparin said people in the Kareila region had often seen footprints of the animal, which they called an abominable snowman.' [4] This is interesting because we seem to have 2 creatures here: one that acts like an animal and the other more human-like.

MOTOR MONSTER: 'Driver Chris Hernandez found a 6ft lizard curled round the engine of his car while checking the steering in Florida'[5] I love these inadvertently mobile animal passenger stories

UNTITLED: 'Mr Philip Nichols, an ex-oil man from Leominster, has died after being knocked over by a sheep.' [6] I am a bit dubious about this one; note the title of the original newspaper. But I thought it was amusing.

And finally: one of my favourites, aberrant terrapins –

SHELL SHOCK: 'Fisherman Roy Peacock hooked a 6 inch terrapin - in a canal. Now Ray, a fireman at Worksop, Notts is keeping his exotic catch in a sink at the fire station while he tries to find it a new home. He said yesterday: "The owner may have set it free because it grew too big for life in a small aquarium."' [7]

1 *Daily Mirror.* July 24th 1992.
2 *The Independent* October 2nd 1992
3 *Daily Telegraph* July 19th 1993
4 *Guardian* April 6th 1992
5 *Daily Mirror* July 8th 1992
6 *Worcester Source* (sic) January 22nd 1987 in *Private Eye* April 3rd 1987
7. *Daily Mirror* (?) May 2nd 1987

The Cure `A Forest`

Come closer and see
See into the trees
Find the girl
While you can
Come closer and see
See into the dark
Just follow your eyes
Just follow your eyes....

1949 turtle hunt log

July 27, 1948

Ora Blue and Charley Wilson, brothers-in-law of Gale Harris, had their fishing rudely interrupted by Oscar, who suddenly surfaced along side of their row boat. They said his back was bigger than the boat and his head the size of a child's. (Oscar Fulk, the original owner of the property, reported seeing the turtle 50 years ago.)

First week of March, 1949

Oscar was seen again, and this time a group of townspeople sought out to capture him. According to Harris and newspaper reports, they actually had the turtle trapped in about ten feet of water off the shore in a trap consisting of stakes and chicken wire but Oscar was too strong and broke out. But at this time a man named Del Winegardner climbed up in a tree and took films of Oscar. Merl Leitch and Dailey Fogle both claim they saw the turtle in the film (though not in person), that it was clearly visible just beneath the water level and was every bit as big as Blue and Wilson claimed. Unfortunately, the film, along with photographs taken by Dailey Fogel, are not available. Mr. Winegardner sold the film four years ago.

March 7, 1949

Whitley County Clerk Charlie White wrote a story in the Columbia City newspaper telling about a turtle "as big as a dining room table top" found near Churubusco. A 'turtling concern.

from Cincinnati, O., has placed a top price of $2,800" on it.

March 8, 1949

The "owner" of the turtle was revealed as Gale Harris. Bill Kellogg, copy editor of The News—Sentinel, gave the turtle the name of "Oscar." The Journal-Gazette named him "The Beast of Busco."

Miss Laura Etz, 23, of the United Press, sent the story out over the wires.

The timing was right and the next day the story of the giant turtle was front page in papers all across the country. An interesting sidelight: James Kirtley, editor of the Churubusco Truth, had the story and was waiting until Thursday, the next edition of the paper—it's a weekly paper—to release it. The story broke on Wednesday and Mr. Kirtley was scooped of the biggest story of his career—right in his own back yard!

March 9, 1949

Whitley County Assessor Lewis Geiger, whose home is near Fulks Lake, said his neighbors were "honest people" and spoke "gospel truth."

March 10, 1949

Kenneth Leitch of the West Side Garage made hooks to catch the turtle.

Bob Shisler rode over the lake looking for the monster from an airplane owned by Carl Sheldon and Ed Keckley.

An editorial in the Journal-Gazette made fun of the story (which was fairly typical of the newspaper accounts) saying they saw the turtle hitchhiking to Fort Wayne; and "spot a catawampus in Franke Park and the turtle boom will collapse in the twinkle of an eye."

March 11, 1949

O. E. Jones, Churubusco, former owner of the farm, said some fellows had told him about a big turtle. "I said it was my Black Angus cow swimming around," he had replied.

Tracks extending 10 to 15 feet were found in the mud.

March 12, 1949

Two hundred people trek out to the Harris farm hoping to catch a glimpse of Oscar and be part of history.

News-Sentinel Photographer Richard Dueter, who had been in the U.S. Navy, suggested fixing a piece of pipe with glass on the bottom to look through the murky water. Dueter and a reporter from the Indianapolis Times said they saw Oscar. Harris saw two different shell patterns.

March 13, 1949

All week-end planes flew overhead and cars moved bumper to bumper, tying up traffic in Churubusco. At night two or three dozen men brought lights to the lake. Ralph Bunn hauled in a "young silo" of a trap, made from quarter inch pipe fastened to a buggy wheel. The men tore up a fence,

SUNDAY, DECEMBER 06, 2009

MUIRHEAD'S MYSTERIES: The Beast of Busco

Today I'm going to focus on Oscar, the Beast of Busco; a giant snapping turtle that inhabited a lake in Churubusco, Indiana in 1949. 'Despite a month-long hunt that briefly gained national attention, the Beast of Busco was never found.' [1].]

July 27, 1948

'Ora Blue and Charley Wilson, brothers-in-law of Gale Harris, had their fishing rudely interrupted by Oscar, who suddenly surfaced along side of their row boat. They said his back was bigger than the boat and his head the size of a child's.' (Oscar Fulk, the original owner of the property, reported seeing the turtle 50 years ago) [2].

'First week of March, 1949. Oscar was seen again, and this time a group of townspeople sought out to capture him. According to Harris and newspaper reports, they actually had the turtle trapped in about ten feet of water off the shore in a trap consisting of stakes and chicken wire but Oscar was too strong and broke out. But at this time a man named Del Winegardner climbed up a tree and took films of Oscar. Merl Leitch and Dally Fogle both claim they saw the turtle in the film (though not in person), that it was clearly visible just beneath the water level and was every bit as big as Blue and Wilson claimed. Unfortunately, the film, along with photographs taken by Dailey Fogel, are not available. Mr Winegardner sold the film four years ago.' [3]

March 11 1949

'O. E. Jones, Churubusco, former owner of the farm, said some fellows had told him about a big turtle. "I said it was my Black Angus cow swimming around," he had replied. Tracks extending 10 to 15 feet were found in the mud.' [4]

End of March

'Four hundred autos an hour snakes there way along the highway to the farm. Letters arrived addressed simply, "Turtle Town, USA." Professional trappers from Tennessee were called in. More divers went under. All these efforts were greatly hindered by the cold and icy March weather. Most observers theorized that Oscar was hibernating in the soft, mushy lake bottom. The town of Syracuse, IN, the site of Lake Wawasee resident and Harris had somehow lured him away into his lake!' [5]

March 19 1949

'The Harris family began selling coffee and hot dogs. Diver Rigsby went down, but the helmet leaked, so he came back up. The search was called off.' [6]

The hunt for Oscar continued through 1949 with various ingenious methods for trying to capture him/her (such as draining the lake) but to not avail. However:

'October 13th 1949 And Oscar didn`t disappoint them [the onlookers]. On a Sunday morning, in plain view of 200 people, he put on his most spectacular show when he leaped out of the water to feast on the live ducks set a top a trap.' [7]

'Finally, October 21st 1949. But this was the last show, for Gale Harris began to run into problems. The soft, mushy bottom presented problems. It started caving in, greatly

reducing the amount of pumping they could do. The pump wore out. Eventually the tractor broke down. To supplement the worn out machinery (and men) a 17 foot crane was brought in. A reporter from a Chicago paper fell into one of the crevices and almost drowned. For two months they pumped and struggled, and in December it was all over. Harris had an attack of appendicitis and when he got out of the hospital, rain had filled up the lake.' [8]

Wikipedia has this to say about cultural impact: 'Oscar`s memory lives on in Churubusco's Turtle Days festival held each June. It includes a parade, carnival and turtle races. A turtle shell labelled "Beast of Busco" hangs in the Two Brothers Restaurant in Decatur, Indiana.' [9]

1.   Wikipedia Beast of Busco http://en.wikipedia.org/wiki/Beast_of_Busco
2.   Churubusco Chamber of Commerce "Oscar The Beast of Busco" p.2
3.   Ibid p.2
4.   Ibid p.2
5.   Ibid. p.3
6.   Ibid. p.4
7.   Ibid p.4
8.   Ibid p.4
9.   Wikipedia op cit.

Devo `Blockhead`

Never leaves a gap
Unfilled
Always pays on time
Always fits the bill
He comes well prepared
Cube top
Squared off
Eight corners
90 degree angles
Flat top
Stares straight ahead
Stock parts
Blockhead
Never tips over
Stands up on his own
He is a blockhead
Thinking man full grown

2 COMMENTS:

Dale Drinnon said...

I live in Indiana. Scuttlebutt about Oscar is that there really was a turtle, probably an alligator snapper, but it was a plant, presumably by the property's owner. The area is outside of the alligator snapping turtle's usual range. There are other ones living there NOW, brought in in memory of the Beast of Busco. No big deal, really.

Lanette said...

Hunting for an alligator snapper is one of the last things I would want to do, have caught too many while fishing for catfish and even the ones a foot in diameter are awful to handle. I would rather handle the last anaconda I had to confiscate.

MONDAY, DECEMBER 07, 2009

MUIRHEAD'S MYSTERIES: Pink unknowns in America

Dear folks, it's time for Muirhead's Mysteries.

Today's edition is entirely devoted to pink unknowns in the U.S. and it is largely down to Mark A. Hall and Anon that we have this information. 'There seems to exist on the North American continent an as-of-yet unidentified species of gigantic amphibian. The amphibian seems to closely resemble the mudpuppy *(Necturus maculosus.)* The mudpuppy is closely related to two species of salamander native to Asia, the giant salamanders *Megalobatrachus davidianus* of China and *M.japonicus*, native to the islands of Japan' [1].

Hall succinctly summarises the above-mentioned phenomena of pink creatures [although the anonymous author of the above article on The Crypto Web does not mention them being pink he/she clearly is referring to the pink cryptids of Scippo Creek, Ohio] thus:'

> Western culture inhibits the discussion of anything that is both strange and pink. Other cultures will be puzzled by our reluctance if not out-and-out incapacity to come to grips with such simple topics. The stumbling block is the familiar allusion to the distorted faculties caused by overindulgence in alcoholic spirits. One who over-drinks is considered inclined to see "pink elephants". By association the report of anything pink and out-of-the ordinary is cause for suspicion and is easily ridiculed. Nature has no respect for our biases. There are unknown animals that are pink and we are going to look at some of the reports here.

> 'In my book Natural Mysteries I discussed the Giant Pink Lizards of Ohio, which appeared to be the larval stage of a giant salamander still unrecognized by establishment scientists. Two centuries ago they were common in the area of Scippo Creek.' [2] The appearance of pink in that case appeared to stem from the albinism of the larval stage.

Anonymous writes:

> 'Early 1800s: Scippo Creek, Ohio. The first report of what may be a giant mudpuppy comes from Scippo Creek in Ohio 9 a tributary of the Scipio River). In the early 1800s, settlers there saw a number of animals, measuring between 6 and 7 feet in length, that were pink in color. These pink water-dwelling lizards had moose-like horns [hardly candidates for salamanders then?] Sometime around 1820, a drought struck the area, drying up numerous streams and creating brush fires which destroyed the local ecology even further. It is generally believed that the animals, whatever they were, were wiped out in these two disasters.' [3]

In 1928 Herbert R. Sass and his wife Marion were boating on Goose Creek near Charleston.

> 'He was on the bow of the boat when they observed a shape moving below the surface of the water. As the boat passed over it Sass extended his oar and managed to lift part of it from the water......It was a bright salmon pink and orange colour.' [4]

In 1968 the American cryptozoologist Ivan T. Sanderson wrote an article in *Argosy* in which he mentions:

> '...how he received a letter from a young woman named Mary Lou Richardson, who said that while hunting with her father she had seen some sort of pinking animal. The creature had a flattened head and a small neck.' [5]

Moving on to 1972: this time Ivan Sanderson himself and his 2nd wife Sabina, saw an 'unknown pinkish critter' in a pond on their land:

'They waded into the pond with a blanket extended between them, set on removing some of the offending growth. Suddenly the blanket parted, torn in half, and something alive showed itself for an instant above the water. What they saw was two feet of something pinkish-orange. It was large and worm-like' [6]

Finally:

'Perhaps there is a pink unknown to account for the report from Vermont. Writing in his newspaper column "Fishy Tales" in the *Rutland Herald*, Charlie Spencer makes reference to a report of a "pink crocodile" in his state. It had been glimpsed in the Tinmouth Channel, which is the name given to the headwaters of the Clarendon River in east-central Vermont' [7]

'Certain peculiarities of the animals in question tally more readily with a mudpuppy explanation, for example the prominent horns of the Scippo Creek animals. In warm, slow moving or stagnant water, the gills of the mudpuppy expand and become much more noticeable. In addition, the largest mudpuppies have been recorded from the southern United States, specifically North and South Carolina - the same general area which has given us several reports of these creatures. The possibility of the existence of such large mudpuppies is an enticing one, although in my opinion, these giant salamanders will probably turn out to be extremely large specimens of *N. maculosus*, rather than a completely new species.' [8]

1.　Anon Giant mudpuppies? http://fortunecity.com/roswell/siren/552/noram_mudpuppy.html [accessed Dec 4th 2009]
2. M.A. Hall Sobering Sights of Pink Unknowns. *Wonders* Dec.1992 p.60 Anon. Giant mudpuppies ?op cit.p.1
3. M..A. Hall Sobering sights op cit p.62
4. Anon. Giant mudpuppies? op cit p.1
5. M.A. Hall. Sobering sights… op cit p.63
6. M.A. Hall op cit p.64
7. K.P.N. Shuker (?) (Unclear from text who author is) *In Search of Prehistoric Survivors*, (1995) in Anon. Giant Mudpuppies?

Led Zeppelin `Stairway to Heaven`

There's a lady who's sure
All that glitters is gold
And she's buying a stairway to heaven
When she gets there she knows
If the stores are all closed
With a word she can get what she came for..

Until we meet again…goodbye!

MUIRHEAD'S MYSTERIES: The dragon of Wantley (a poem)

R E L I Q U E S
OF
ANCIENT ENGLISH POETRY:
CONSISTING OF
Old Heroic BALLADS, SONGS, and other
PIECES of our earlier POETS,
Together with some few of later Date.
THE THIRD EDITION.
VOLUME THE THIRD.

DURAT OPUS VATUM

L O N D O N:
Printed for J. DODSLEY in Pall-Mall.
M.DCC.LXXV.

Dear folks,

I have in my possession a poem, or rather doggerel, about the Dragon of Wantley, which I found in *Folk Tales of the British Isles*, Folio Society (1985). It is far too long to reproduce in its entirety here so I reproduce 7 verses. But first of all, a little bit about the dragon him/her/itself:

'The Dragon of Wantley is a 17th century satirical verse parody about a dragon and a brave knight. It was included in Thomas Percy's 1767 Reliques of Ancient Poetry.

'The poem is a parody of medieval romances and satirizes [sic] a local churchman. In the poem, a dragon appears in Yorkshire and eats children and cattle. The knight More of More Hall battles the dragon and kills it. The Wantley of the poem is Wharncliffe, as the dragon lived in a cave on Wharncliffe Crags, five miles to north of Sheffield, South Yorkshire, Sir Francis Wortley, the diocese ecclestiastic, and the parishioners of Wharncliffe had a disagreement on tithing and how much the parish owed (under the law of "First Fruits"), so the poem makes him a dragon. More of More Hall was a lawyer who brought a suit against Wortley and succeeded, giving the parishioners relief. Thus, this parody romance satirizes Wortley. The author of the poem is unknown.' [1]

The poem was transformed into an opera in 1737 attacking Robert Walpole's taxation policies. Owen Wister wrote a novel, *The Dragon of Wantley*, in 1892 'It is a romantic story set at Christmastime in the early 13th century. The book was a surprise success, going through four editions over the next ten years.' [2]

THE DRAGON OF WANTLEY

Old stories tell, how Hercules
A dragon slew at Lerna,
With seven heads, and fourteen eyes,
To see and well discern-a:
But he had a club, this dragon to drub,
Or he had ne'er done it, I warrant ye;
But More of More Hall, with nothing at all,

He slew the dragon of Wantley.

This dragon had two furious wings,
Each one upon each shoulder;
With a sting in his tail, as long as a flail,
Which made him bolder and bolder.
He had long claws, and in his jaws
Four and forty teeth of iron;
With a hide as tough as any buff,
Which did him round environ.

Some say, this dragon was a witch;
Some say, he was a devil,
For from his nose a smoke arose,
And with it burning snivel,
Which he cast off, when he did cough,
In a well that he did stand by;
Which made it look just like a brook
Running with burning brandy.

Hard by a furious knight there dwelt,
Of whom all towns did ring,
For he could wrestle, play at quarter-staff, kick, cuff and huff,
Call son of a whore, do anything more;
By the tail and the mane, with his hands twain,
He swung a horse till he was dead;
And that which is stranger, he for very anger
Ate him all up but his head.

To see this fight, all people then
Got up on trees and houses,
On churches some, and chimneys too,
But these put on their trousers,
Not to spoil their hose. As soon as he rose,
To make him strong and mighty,
He drank by the tale six pots of ale,
And a quart of aqua-vitae.

At length the hard earth began to quake,
The dragon gave him a knock,
Which made him to reel, and straightaway he thought
To lift him as high as a rock,
And thence let him fall. But More of More Hall
Like a valiant son of Mars,
As he came like a lout, so he turned him about,

and hit him a kick on the arse.

'Murder, murder!' the dragon cried,
'Alack, alack for grief!'
Had you but missed that place, you could
Have done me no mischief.'
Then his head he shaked, trembled and quaked,
And down he laid and cried;
First on one knee, then on back tumbled he,
So groaned, kicked, shat, and died. (3)

1. Wikipedia. Dragon of Wantley. http://wikipedia.org/wiki/Dragon_of_Wantley [accessed Dec7th 2009]
2. Ibid
3. Author unknown.The Dragon of Wantley. *Folk Tales of The British Isles*. Folio Society 91985) pp139-144

Rich. Sorry no song lyrics today due to disorganized brain cells.

WEDNESDAY, DECEMBER 09, 2009

MUIRHEAD'S MYSTERIES: "Giant Spiders" and a Yorkshire cave

Today I take a look at a story that appeared in the British media last week and caught my eye. For those of you who missed it, it's about a colony of spiders from a cave in the Yorkshire Dales; more specifically, Chapel Fell cave. If you are an arachnophobe look away now.

'Huge UK cave spiders 'sent' home. A colony of huge cave spiders is finally heading home after 10 years. The spiders have been squatting in a disused building in the Yorkshire Dales after escaping from a nearby cave, hidden in scientists' equipment. Volunteers and staff from the National Trust's Malham Tarn estate in North Yorkshire are now transporting the spiders back to their natural home. Measuring seven centimetres across, the cave spiders are amongst the largest spiders found in the UK. "The time has come for the cave spiders to be relocated back to their natural homes," says Martin Davies, National Trust property manager for the Yorkshire Dales in the UK. The old house is due to be renovated for use by visiting schoolchildren and walkers, Mr Davies explains.' [1]

The Guardian commented on the same day:

'More than 150 of the cave-dwelling species *Meta menardi* and *Meta bourneti* are being collected individually and taken in plastic bags to a pothole in the Yorkshire Dales. The

journey returns them to their original home, which they left - unnoticed - 10 years ago on the clothes and equipment of a party of university scientists.

'"Exceptionally for cave spiders, which have adapted to live underground, they resettled and flourished in a derelict orchid house which the scientists were using as their base... They clearly took to it [the orchard house] immediately, although we only realised that they were there a couple of years ago," said Mike Collins of the National Trust, which runs the former mansion of Malham Tarn House, North Yorkshire, as a field centre. The big move follows the trust`s decision to convert the orchid house into a classroom, with facilities such as warmth and light, which the spiders loathe. Familiar from their cave roof cocoons, from which they sally to find prey, they measure up to 8cm (3.1in) across and will nip if repeatedly provoked".

[So a bit bigger than the BBC`s measurement.]

'...Visitors will be able to learn about spiders generally, the Meta species and the story of Malham migration, while the descendants of the original travellers get on with life in the dark back at Chapel Fell' [2]

I visited Malham Tarn House in the 1980s on a geography field trip; it's a lovely place.

1. J. Bourton. Huge UK cave spiders` sent home. BBC Earth News Dec.5th 2009.http://news.bbc.co.uk/earth/hi/earth_news/ newisd_8393000/8393757.stm [accessed Dec.8th 2009]
2. M. Wainwright. Spiders to be sent back to their cave 10 years after their great escape. Guardian December 5th 2009

Devo `Too Much Paranoias`

Think I got your dial tone
Think I got Billy Baxter's Bone
Think I got a bubble sac
Think I got a Big Mac Attack
Hold the pickles hold the lettuce
Special orders don't upset us
All we ask is that you let us
Serve it your way
Too much paranoias
My momma's afraid to tell me
The things she's afraid of
I been dipped in double meaning
I been stuck with static cling
Think I got a rupto-pac
Think I got a big mac attack

THURSDAY, DECEMBER 10, 2009

MUIRHEAD'S MYSTERIES: The Bosom Serpent

Dear folks,

This information on the rather grotesque-sounding `bosom serpent` is taken from Dr Jan Bondeson's article `The bosom serpent` in the *Journal of The Royal Society of Medicine* vol. 91 August 1998. Dr Bondeson kindly sent me a large amount of material on the basilisk and also a selection of material on medical curiosities after Weird Weekend 2009.

According to one of the old Viking annals, the Flato Book, King Harald Hårdråde of Norway once visited the nobleman Halldor, whose daughter had been very ill. Fever, increasing abdominal girth and an unquenchable thirst were the major symptoms. The old woman gossiped about her being pregnant, but the young lady denied this with great vehemence. Since she was steadily declining, the King was consulted. His diagnosis was that she had accidentally swallowed the spawn of a serpent when she drank water, and the reptile had grown within her....

'The belief in living snakes, frogs, lizards and other animals as parasites within the human gastrointestinal tract is of considerable antiquity. Already in ancient Egyptian, Assyrian and Babylonian manuscripts there is mention of a `colic-snake` as a cause of painful stomach cramps. In De Morbis Vulgaribus, Hippocrates describes the case of a youth who had drunk a great quantity of strong wine. When he passed out on the ground, a snake slithered down his throat and caused his death from an apoplectic seizure....

'The bosom serpent also plays a part in one of the legends about the medical saints Cosmas and Damien, who were martyred in AD 283. A poor peasant was tortured day and night by a large snake that had crawled down his throat while he slept. No doctor could help him, but when he stepped into Cosmas and Damien`s church the serpent slithered up his throat with great haste...

'...In 1618, the Prebendary of Strasbourg, Dr Melchior Sebizius, reported another famous case. A 17-year-old youth had consulted him for stomach pains, melancholia, flatulence and epileptic seizures but he had been unable to diagnose the ailment. Some weeks later, the youth was found sitting dead in a bog-house; beneath the seat, a large snake was crawling about. In a thesis, illustrated with a figure of the snake...Melchior Sebizius concluded that the unhappy youth`s afflictions could all be explained by the presence of the snake in the stomach for an extended period of time, and that the strain of expelling it from his body had induced a fatal apoplectic fit....

'In 1694,the 12-year-old son of Pastor Zacharias Döderlein, of Berolzheim in southern Germany, was taken severely ill. After several fits and attacks of abdominal cramps, he vomited numerous insects, and later also 21 newts, 4 frogs and some toads...These uncanny happenings soon attracted notice among the clergy: many German parsons came to visit the house of their stricken colleague and their diagnosis was that the boy was possessed by the Devil; what particularly impressed them was that when the boy was led to take some fresh air near a pond with croaking frogs, his stomach - frogs croaked loudly in reply...In the late 18th century, most of the leading biologists, including Linnaeus, Buffon and Blumenbach, favoured the notion that snakes and frogs could live as parasites in the human gastrointestinal tract....

'In 1850, Professor Arnold Adolph Berthold, of Göttingen, published a monograph aimed at solving the burning question of the existence of living amphibians as parasites in the human stomach. He had noted that almost every German pathological museum of repute contained some snake, frog or newt which had allegedly been vomited by some patient after living for years within the human body. Berthold obtained permission to dissect several of these specimens and all had partly digested insects in their stomachs - a strong indication that they had been deliberately swallowed shortly before being vomited....

'Early in 1916, a remarkable story appeared in several English newspapers. A woman had swallowed frogspawn, which had developed inside her into a large frog. She was taken to Stroud Hospital...but the doctors were unable to operate since the animal moved about too fast. The woman was in such agony that the baffled medical men wrote to King George V in order to get his permission to kill her with poison, but his Majesty refused this plea.

'Another canard went the rounds of the newspapers in 1987. An 11 year old girl in Baku had got a 26-inch semi poisonous Caucasian cat snake in her stomach, which had slithered down her throat while she slept. The clever doctors managed to flush it out with a stomach-pump. The Scotsman added a further frisson by adding that she was still in hospital, being treated for the enraged reptile's bites in the stomach wall.' (1)

Dr Bondeson concludes that the bosom serpent was likely psychosomatic or a parasite.

1.   J. Bondeson. The bosom serpent. *Journal of the Royal Society of Medicine*. Vol. 91 August 1998. pp.442-447

FRIDAY, DECEMBER 11, 2009

MUIRHEAD'S MYSTERIES: Tanzanian chameleon

Today, Friday December 11th, Muirhead's Mysteries features snakes yet again, but this time in a far less gruesome way. Indeed, in a positive way. The headline to this piece, from guardian.co.uk says it all:

'Snake spits out new species of chameleon at scientist's feet. `

Latest find in natural world was result of reptile coughing up lizard as conservationist studied monkeys in jungle.

"It was so nearly known as dinner. Instead, a small and not terribly impressive chameleon has become the newest discovery of the natural world, after a startled Tanzanian snake spat a still undigested specimen at the feet of a British scientist, who identified it as a previously unknown species. Dr Andrew Marshall, a conservationist from York University, was surveying monkeys in the Magombera forest in Tanzania, when he stumbled across a twig snake which, frightened, coughed up the chameleon and fled. Though a colleague persuaded him not to touch it because of the risk from venom, Marshall suspected it might be a new species, and took a photograph to send to colleagues, who confirmed his suspicions."

'"The thing is, colour isn't the best thing for telling chameleons apart, since they can change colour for camouflage. They are usually identified based on the patterning and shape of the head, and the arrangement of scales. In this case it's the bulge of the scales on its nose."

'Happily for Marshall, shortly afterwards he spotted a second chameleon, this time alive, and was able to photograph it. The two creatures were found about six miles apart, which he believes may be the full extent of the area colonised by the extremely rare species.

Though he found the specimen in 2005, his paper on the discovery, published this week, puts the find formally on record. "It takes quite a long time to convince the authorities that you have a new species." He said.

'Had Marshall hoped it might be named after him? "Oh crumbs, no. The thing is, if you work in an area of conservation importance and you give a species the name of that area it can really highlight that area. By giving it the name Magombera it raises the importance of the forest."' [1]

1. Snake spits out new species of chameleon at scientist's feet. http://www.guardian.co.uk/world/2009/nov/23/new-chameleon-species-magombera-ta

Devo 'Space Junk'

She was walking all alone
Down the street in the alley
Her name was Sally
She never saw it
When she was hit by space junk

In New York Miami beach
Heavy metal fell in Cuba
Angola Saudi Arabia
On Xmas eve said norad
A Soviet Sputnik hit Africa
India Venezuela (in Texas Kansas)

It's falling fast Peru too
It keeps coming
And now I'm mad about space junk
I'm all burned out about space junk
Oooh walk & talk about space junk
It smashed my baby's head
And now my Sally's dead

SATURDAY, DECEMBER 12, 2009

MUIRHEAD'S MYSTERIES: Another giant turtle in the U.S.A.

A few days ago you may remember I presented information about a very large turtle in a U. S. lake, namely, The Beast of Busco. Well, today I turn to a similar turtle, this time in a lake in Illinois, namely Island Lake quite near Chicago. This latter one was even thought to be a killer

I found this brief story in *Omni* magazine, the now defunct U. S. science fiction, fantasy and paranormal magazine. Unfortunately I do not know the exact date; probably some time in the mid to late 1990s, though it was in August. There is a link on the Web to some of the archives at:

http://web.archive.org/web/*/http://www.omnimag.com/

'...the links past Jul 21, 2003 will go to *Penthouse* magazine instead, but most of them are web-archived *Omni* pages, including chats, short stories, and articles.' [1]

'OMNI was a science magazine and science fiction magazine published in the U. S. A. It contained articles on science fact and short works of science fiction. The first issue was published in October 1978, the last in Winter 1995, with an internet version lasting until 1998.' [2]

The story is as follows: 'MOBY TURTLE. Move over Nessie. Recent reports tell of a giant killer turtle living in Island Lake, Illinois, some 40 miles north-west of Chicago. It all started about a year ago, when resident Liz Herman spotted two large snapping turtles mating in the water just beyond her back-yard sea wall.

'"These turtles were huge, maybe three feet long" says Herman. "My husband Kevin didn`t believe me until he saw my photos. Then he showed them around town." "Soon rumors [sic] were flying." says Georgine Cooper, a town trustee. "The turtle was said to be anywhere from the size of a small pickup truck to an 18-wheel semi." "Our Water Department supervisor, Neil De Young, tried to catch the turtle," adds local police chief Eugene Bach." All he got for his trouble was bent hooks."

'Bach refuses to fuel the killer-turtle rumours. In fact, he thinks the turtle may be an alligator snapper released into the lake some years ago by a local boy." Alligator snappers, he says, "can grow to more than 60 pounds and can be 25 inches long. Listen," he adds, "I could tell you that when kids hang their feet over the pier they get chewed up, but that wouldn`t be true." [3]

There is a headline in the text which says, in capital letters: 'The Giant Turtle, Referred to as a man-eating terror, has generated national tabloid headlines and has even been profiled on local Chicago news stations.' [4]

1. *Omni* (magazine) http://en.wikipedia.org/wiki/Omni_(magazine)
2. Ibid
3. Moby Turtle. *Omni.* August?p.72
4. Ibid.p.72

Neil Young `Heart of Gold`

I've been to Redwood
I've been to Hollywood

I've crossed the ocean
for a heart of gold
I've been in my mind
It's such a fine line
That keeps me searching for a heart of gold
And I'm getting old

3 COMMENTS:

Retrieverman said...
The word "Rumors" is not spelled incorrectly in American English. Because this was published in American English, it is spelled correctly, and it does not need (sic) in parentheses beside it.

Color is how we spell colour.

Realize is how we spell realise.

My browser underline all misspelled words, and when I typed those words with their British spelling, they were underlined in red.

Our language is standardized and distinct, and although our country is run by imperialists who think the world is theirs, our language is as it is.

Retrieverman said...
I'll just add that I always call this organization (or organisation) the See Eff Zed, even though in my language it is See Eff Zee.

I think that we should accept the two different versions of English.

sammywise said...
When I was a teen I found two very large "turtle scrappings" on a small tributary of the Ottawa River in Toledo.
The claws of the feet were clearly visible. One set was somewhat smaller than the other. The largest was wider than my arm reach and the smaller one was a few inches less.

SUNDAY, DECEMBER 13, 2009

MUIRHEAD'S MYSTERIES: Large lizards in early 19th century New Zealand

Today I'm roving around the world again to New Zealand or more specifically, early nineteenth-century New Zealand. I quote extensively from the *Transactions and Proceedings of the New Zealand Institute* vol.7 1874. A long time ago an American cryptozoologist sent me this and it has been lying dormant (rather like a mythical beast, no less) in my files until now.

`On the Disappearance of the larger Kinds of Lizard from North Canterbury. By Rev. J. W. Stack.

"The absence of living specimens, coupled with the absence of all traces of recent remains, would render the task of proving that the large lizards existed till quite lately in this part of the country very difficult but for the fact that there are many Maoris still living who have not only seen but handled and even eaten them. To prevent the knowledge of an interesting zoological fact being lost I have written down the statements of such of the natives whose testimony seemed most worthy of credit. They are persons whose names appear in the earliest records of the colony as leading members of the native community, and therefore from their age may be considered competent to give evidence upon matters of fact which occurred under their observation forty or fifty years ago.

'The following is a summary of the statements made by Te Aika,Te Uki,Iwikau, and Te ata o Tu:-

'Unu ngara or ngarara burrows were frequently met with on the plains. They were plentiful in the manuka scrub extending from the banks of the Waimakariri past the present site of Eyreton westwards towards the ranges, and at Waitui,between the Hurunui and Waiau rivers. The ngara was darker in colour than the ruatara. They varied in size from two to three feet in length, and ten to twenty inches in girth; along the back from the nape of the neck to the tail was a serrated crest. The mouth was full of teeth, some grew large and caused the upper lip to project. These when taken from the jaw were three or four inches long, and half an inch at the base; when split in two and polished they were prized as mat pins.

'A ngara known as Te iha was kept a long time at Kaiapoi. It was fed on small birds and prepared fern-root. It was very gentle and liked being stroked, uttering at the time a gutteral sound expressive of pleasure. When it made this noise at any other time it was indicative that it wanted food and water.

'Besides the kind frequenting the manuka scrub there was a smaller ngarara, about eighteen inches long, found in the strams. Horomona Iwaiku was eeling some time before the fall of Kaiapoi at Orawhata, a stream which rises near Riccarton and falls into the Waimakariri. After having caught a great many eels, which he killed with a billet of wood, he was terrified by the cries of one he was in the act of killing; though very frightened, he continued to strike till the sound ceased. On examination he found it was a ngara; becoming emboldened he lit a fire, and cooked and ate it. The natives attribute the disappearance of the large ngara to the introduction of cats and to frequent fires. The Norwegian rat has probably a hand too in the extinction of these reptiles." [1]

1. Rev. J.W. Stack.On the Disappearance of the larger Kinds of Lizard from North Canterbury. *Transactions and Proceedings of the New Zealand Institute.*vol.7 1874 pp 295-296.

Richard. Sorry, I forgot about lyrics. Try listening to the *Ballad of Frankie Lee and Judas Priest* by Bob Dylan.

EDITOR'S NOTE: I would go further and recommend that if you have never done so that you listen to the entire *John Wesley Harding* album. It is my favourite Dylan album, and the song cited by Richard is my favourite track.

1 COMMENT:

shiva said...
Interesting. Sounds like another species of sphenodontian, like the tuatara, rather than any of the groups of lizards (geckos and skinks) found in New Zealand, although it also sounds reminiscent of some iguanas, such as the Galapagos marine iguana.

From the Wikipedia article on the tuatara: "A third, extinct species of Sphenodon was identified in November 1885 by William Colenso, who was sent an incomplete sub-fossil specimen from a local coal mine. Colenso named the new species S. diversum." There's a PDF of the source. Also, there's this extinct giant gecko, which seems like something of a cryptid itself (but has a completely different Maori name, suggesting it's not the same thing): http://en.wikipedia.org/wiki/Kawekaweau

TUESDAY, DECEMBER 15, 2009

RICHARD MUIRHEAD: The `Kawekaweau` another giant lizard in New Zealand. Part one.

According to The Dear Leader (Kim Jong Il - Jon Downes to you living outside North Korea), The Great Leader (Kim Il Sung Richard Freeman) has already commented upon the Kawekaweau, a giant gecko, which may still inhabit the Gisborne Region of New Zealand, in his piece on cryptids in museums. However, I have a bit of information on the Kawekaweau that may not have seen widespread publicity. Since the spring of 1990, when news emerged that a museum in Marseille was transferring its specimen of the kawekaweau to New Zealand, the track of this cryptid has gone cold as far as I am aware.

However, I have today emailed the National Museum of New Zealand in Wellington to ask if Tony Whitaker has found out any more information on the status of the Kawekaweau *(Hoplodactylus delcourti)* especially in the Tolaga Bay area of East Cape. Anthony H. Whitaker was located at the Natural History Unit, National Museum, Wellington,

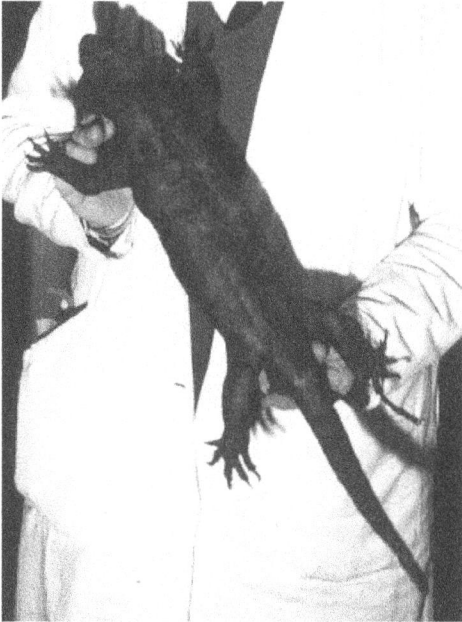

in 1990. If I get a reply I will try and remember to let you know.

According to Bauer and Russell:

'Examination by Aaron M. Bauer revealed that the specimen represents a new species of the genus Hoplodactylus, until now, known only from New Zealand. Subsequent inquiry has revealed a possible connection between the specimen and the kawekaweau of Maori legend. The huge size of the specimen extends our conception of the morphological extremes attainable by gekkonid lizards. It further suggests questions regarding conservation of native reptiles, and the investigation of rare or recently extinct taxa.' [1]

The authors later go on to comment on the origin of the specimen. Referring to the Musee d'Histoire Naturelle de Marseille: 'The collection locality of the specimen is in doubt. Although no data exist, the morphology of the specimen limits the area of possible origin, as it represents a member of a limited radiation of the subfamily Diplodactylinae...in the southwest Pacific Ocean. The first possibility is that the animal originated in the French island territory of New Caledonia....An alternative hypothesis of a New Zealand origin for *Hoplodactylus delcourti* would be consistent with current ideas regarding the distribution of this genus, and would not be inconsistent with the history of many areas of the North Island, particularly in the Bay of Islands...The contention that the specimen is from New Zealand is also supported by some historical and anecdotal records from both Maori and European sources. A number of the older descriptions of the lizards of Maori legend are suggestive of the new species. The most valuable information comes from Mair (1873-2) who commented on "the existence of a large forest lizard", called by the Maoris kaweau." He continues, "In 1870 an Urewera chief killed one under the loose bark of a dead retain the Waimana Valley, he described it to me as being about two feet long and as thick as a man's wrist; colour brown, striped longitudinally with dull red."'[3]

Due to the quantity of source material available, this blog will continue on Wednesday 16th Dec.

1. A.M. Bauer and A.P. Russell. *Hoplodactylus delcourti n.sp* (Reptilia:Gekkonidae) the largest known gecko. *New Zealand Journal of Zoology*, 1986 vol 13. p.141
2. W.G.Mair.1873 Notes on Rurima Rocks.*Transactions of the New Zealand Institute* 5:151-153
3. Bauer and Russell.Ibid.p.146

WEDNESDAY, DECEMBER 16, 2009

MUIRHEAD'S MYSTERIES: The Kawekaweau part two

Today I am continuing my look at the kawekaweau of North Island, New Zealand. It turns out one of my correspondents might be interested in cryptozoology. Or at least I may have told him about Cryptozoology Online and he looked at my first, pre-kawekaweau posting on large New Zealand lizards. To be honest I can't recall if I told him or not. He is Raymond Coory, technician at the Museum of New Zealand, and he has given me the contact details of some other individuals interested in the kawekaweau.

I will now continue on from yesterday's blog:

'Current status. Although no living *Hoplodactylus delcourti* are known, we cannot deny the possibility that populations of this species still exist. If the species is indeed the kawekaweau of Maori legend, annecdotal sources suggest that the northern North Island may be a region likely to support surviving populations.' [1]

A. P. Russell and A. M. Bauer comment:

'*Hoplodactylus delcourti*, with a snout-vent length of 370mm increases the recorded for the family by 54% and far exceeds the maximum dimensions of its congeners. The largest previously known specimen of *Hoplodactylus* was a specimen of H. duvaucelii, with a snout-vent length of 160mm... This new taxon represents an increase of 131.3% over this dimension' [2]

'Kawekaweau (pronounced cah-way-cah-way-ow, but also known as kaweau or koeau) were reported from widespread localities in the northern half of the North Island. The animal was variously described as being amphibious, a ground dweller, a tree-dweller, or even being able to fly! The most often repeated description was of a lizard about two feet long that was arboreal...Then, in 1979, Alain Delcourt, a curator at the Museum d'Histoire Naturelle in Marseille, discovered in the museum's zoological collections a mounted specimen of a huge gecko..

'When and how the Marseille specimen got to France is unknown. French expeditions have visited New Zealand since the earliest times of European exploration. There have been French missions and French settlements. Marseille was regarded as the "gateway to the Orient", and was the home port for most voyages to this part of the world. Biological specimens from collectors and curios from seamen would pass through Marseilles on their way into France. The huge gecko was probably acquired by the museum between 1833 and 1869, a period for which all the records have been lost.' [3]

I have a report titled *Large Lizard Sightings in the Gisborne Region: Report on a National Museum Investigation 7-11 April 1990* by A. H. Whitaker and B. W. Thomas. This was 1 month after the stuffed 'kawekaweau' arrived at the National Museum in Wellington. 'Four of the sightings by four different observers are not so easily explained. These people are largely unknown to each other yet on separate occasions over a 30 year period saw what are described as remarkably similar animals. More remarkable is that these observations were all within a few kilometres of each other on a short stretch of Anaura Rd, north of Tolaga Bay. Possible explanations are that...there is a hitherto unrecorded indigenous lizard present.' [4]

On July 15th 1994, I received a letter from Chris Paulin of the Natural Environment Section of the Museum of Wellington saying:

'Thank you for your letter regarding lizard and moa sightings. There have been no large lizard sightings in the last twelve months. The moa sightings reported in the press originated from a local hotelier and were based on some very blurred photographs of a red deer.' [5]

Raymond Coory, mentioned at the beginning of today's blog, says the kawekaweau is covered in the *International Society of Cryptozoology Newsletter* vol 7 no 1 and vol 9 no 4. He concludes:

'Did it really live here and could it possibly still be alive? We have a constant stream of sub fossil and fossil bones coming through our research departments, and among the millions of bones there have been many vertebrate fossils of birds, lizards and frogs but nothing from a big gecko. Also New Zealand is intensively tramped and there have been no credible sightings or photographs. An attempt was made to get DNA from the French specimen in 1990, but failed. The technology is vastly more sensitive now so another go would probably yield results. I don't know if this has already been attempted, but the results would be very interesting. [6]

I will be e-mailing the people Raymond Coory gave me the details of before the New Year. If anyone has any specific questions they want me to ask about this creature please contact me.

1. A.M. Bauer and A.P. Russell Hoplodactylus delcourti n.sp.
   Reptilia:Gekkonidae),the
   largest known gecko. *New Zealand Journal of Zoology* 1986. p.147
2. A.P. Russell and A.M. Bauer The Giant Gecko *Hoplodactylus delcourti* and Its
   Relations to Gigantism and Insular Endemism in the Gekkonidae. *Bull.Chicago Herp.Soc.* 26 (2):p.26.1991
3. T. Whitaker Kawekaweau-mtyh or monster? *New Zealand Geographic* 6 p13 1990
4. A.H. Whitaker and B.W. Thomas. *Large Lizard Sightings in the Gisborne Region: Report on a National Museum Investigation* 7-11 April 1990 p.1
5. Letter from C. Paulin to Richard Muirhead July 15th 1994
6. E-mail from Raymond Coory to Richard Muirhead December 15th 2009

Bob Dylan `Ballad of a Thin Man`

You walk into the room
With a pencil in your hand

You see someone naked
And you say "Who is that man?"
You try so hard
but you don't understand
just what you'll say
When you get home

1 COMMENT:

Bruce Spittle said...
Further details on the hotelier's moa sighting claim referred to in reference 5 are given in a new 3-volume book Moa Sightings, published on 1 January 2010, and available from www.moasightings.com
Bruce Spittle

THURSDAY, DECEMBER 17, 2009

MUIRHEAD'S MYSTERIES: Strange creatures in an Israeli cave

Dear cryptofolk,

It being nearly Christmas I thought I'd introduce an item from Israel, namely the new fauna, both (possibly) extinct and still living in the Ayalon Cave, on Israel's coastal plain near Ramle, between Jerusalem and Tel Aviv. This unique ecosystem was discovered in about April 2006. Another interesting occurrence was the sighting of a 'mermaid' in the eastern Mediterranean off Israel's coast in or very near August 2009.

A Department of Media Relations (part of The Hebrew University of Jerusalem) website press release titled 'Unique Underground Ecosystem Revealed by Hebrew University Researchers Uncovers Eight Previously Unknown Species' gives the following information:

> 'Discovery of eight previously unknown, ancient animal species within a new and unique underground ecosystem in Israel was revealed today by Hebrew University of Jerusalem researchers.
>
> 'In a press conference on the Mt. Scopus campus of the Hebrew University, the researchers said the discovery came about when a small opening was found, leading to a cave extending to a depth of 100 meters [sic] beneath the surface of a quarry in the vicinity of Ramle, between Jerusalem and Tel Aviv. The quarry is operated by cement manufacturer Nesher Industries...
>
> 'The invertebrate animals found in the cave-four seawater and freshwater crustaceans and four terrestrial species-are related to but different from other, similar life forms

known to scientists. The species have been sent to biological experts in both Israel and abroad for further analysis and dating. It is estimated that these species are millions of years old. Also found in the cave were bacteria that serve as the basic food source in the ecosystem...

'The animals found there were all discovered live, except for a blind species of scorpion, although Dr Dimentman is certain that live scorpions will be discovered in further explorations and also probably an animal or animals which feed on the scorpions...

[Dr Dimentman was part of the team from the Hebrew University Alexander Silberman Institute of Life Sciences, which discovered the fauna]

'Among the interesting features of the discoveries thus far in the cave is that two of the crustaceans are seawater species and two others are of types found in fresh or brackish water. This can provide insights into events occurring millions of years ago regarding the history of ancient bodies of water in the region.' [1]

I have a paper from 2007 that gives more information on the troglodite scorpion. Most of it is too technical for me to understand so I just quote from the Abstract and Introduction:

'Abstract. A new eyeless troglobite scorpion, Akrav israchanani n.sp., is described from inside karstic voids in Israel that form a completely isolated, old ecosystem with living populations of blind crustaceans and pseudo scorpions. The scorpions, of which no live specimen has yet been collected, prove to differ from all other scorpions and are placed in a new family, Akravidae. The possibility is addressed that the subterranean Akravidae are a relict of an old circum-tropical pattern of distribution that differs from the present temperate location of Israel.'

According to the Introduction: 'No live scorpions have as yet been detected, only their empty carcasses. These desiccated but not fossilized [sic] cuticular remains which retain their bright fluorescence under UV light were found firmly attached to rocks at various levels corresponding to the levels attained by the rise and fall of the underground water inside the voids. No traces of any of the scorpions prey animals have yet been found. Could their disappearance have caused the presumed extinction of the scorpions?' [2]

On December 15th 2009 I received the following e-mail from Prof. Amos Frumkin of the Geography Department at The Hebrew University:

'Dear Richard,

We're still working on the Ayalon Cave... No new discoveries of animals been made since 2006. Lots of other new caves been discovered elsewhere in Israel since the Ayalon caves were discovered, albeit not with new fauna. I don't know of any organization [sic] in Israel that keep records of any new animal species that are discovered in Israel since 1948, cheers Amos. '[3]

That's all for now. See you tomorrow. And we have lyrics!

1. Dept.of Media Relations.The Hebrew University of Jerusalem press release.http://www.huji.ac.il/cgi-bin/dovrut_search_eng.pl?mesge114907691205976587
2. G.Levy.The first triolobite scorpion from Israel and a new chactoid family (Arachnida:Scorpiones)Zoology in the Middle East 40,2007:p.91
3. E-mail from Prof.Amos Frumkin. Dec.15th 2009.

The B52s -`Give Me Back My Man`

She cuts her hair and calls his name
Wishin' everything could be the same
like when she had him

I'll give you fish, I'll give you candy
I'll give you everything I have in my
head

Walking out of Korvettes, package in her hand.
Motions to all the seabirds, throws divinity on the sand

FRIDAY, DECEMBER 18, 2009

MUIRHEAD'S MYSTERIES: Dragons of Sweden

Hello again, cryptophiles.

Today I present a summary of an article in *Fate* magazine concerning `The Dragons of Sweden`. Unfortunately I do not know when it was published; my apologies. The article is by Sven Rosen. Unfortunately the quality of the photocopy I have is poor so there are parts I cannot transcribe.

'In the mid-19th century Swedish newspapers published a number of reports in which people described encounters with dragons or giant snakes. Fascinated by the stories, the pioneering Swedish folklorist and scientist Gunnar Olof Hyltén-Cavallius decided to investigate. From his studies he knew that such creatures, called lindorms, were widely known in Nordic Europe. Lindorms figure in Scandinavian mythology, in folk songs and tales and in the dragonesque style of the Viking Age. Throughout history descriptions of the creatures` appearance were remarkably consistent.Hyltén-Cavallius noted, and now it appeared that specimens of this rare monster still roamed the Swedish countryside. So

he set out to find a dragon....' [1].

This is Hyltén-Cavallius' summary of his findings: 'In Värend - and probably in other parts of Sweden as well - a species of giant snakes, called dragons or lindorms, continues to exist. Usually the lindorm is about 10 feet long but specimens of 18 feet or 20 feet have been observed. His body is as thick as a man's thigh; his color [sic] is black with a yellow-flamed belly. Old specimens wear on their neck an integument of long hair or scales, frequently likened to a horse's mane. He has a flat, round or squared head, a divided tongue, and a mouth full of white, shining teeth. His eyes are large and saucer-shaped with a frightfully wild and sparkling stare. His tail is short and stubby and the general shape of the creature is heavy and unwieldy...' [2].

'The dragon resides in dens and piles of stones, in wild and desolate places (?) marshlands, swamps or lakes. He often (?) been seen in the rocky mountains and forests east of Lake ⬛snen. He has also been observed swimming in the lakes of (?) Rottnen and Helgasjön. Usually he (?) keeps his head two feet above water...he moves onward with the same kind of winding throwings as an ordinary grass snake...'[3].

'An encounter with a lindorm consists of one or more of these elements;

1. Observation. The witness happens to see a dragon. The encounter is peaceful although the witness often is deeply affected by the snake's appearance.

2. Pursuit. The snake, which moves quickly in an upright position and often snorts like a horse, chases the witness. (Of Hyltén-Cavallius' 48 cases only two deal with this type of event.)

3. Combat. The witness fights the snake (12 cases)

4. Appearance with other, smaller snakes (four cases)' [4]

Later the author describes the foul stench the decomposing carcass makes:

'The snake's breath and the stench that he emitted after his death was so poisonous that I for eight days thereafter suffered from the after effects of it, and I and my companion could hardly come near to dead snake's body [Case five.] The stench of the carcass, as well as the smell of the venomous liquid that he [the lindorm] vomited forth on me during the struggle was unbearable, and I was, because of that, confined to bed for three days thereafter, and was badly ill. [Case 23]....' [5].

'Descriptions of the lindorm appear frequently in Scandanavian literature from the 13th-Century Eddies on. Olaus Magnus, in his *Historia de Gentibus Septentrionalibus* (1555), mentions in particular the dragon's hard, scaly body. Sigurd of the Eddas and Siegfried of the Nibelungenlied found that a dragon's weak point is its belly; so did Daniel Nilsson of Värend in Hyltén-Cavallius' case four' [6].

The author concludes with considering a psychological explanation:

'One major problem with this psychological interpretation is that 24 of Hyltén-Cavallius' cases involve more than one witness. Many of the 31 additional cases with which I am familiar also had multiple witnesses. One can speak of "collective hallucinations" without effectively explaining anything. Still it is possible, in my opinion, that the parallels between the monster stories from Värend and the characteristic features of hallucinations are not entirely coincidental.' [7]

There is information in *Lake Monster Traditions* by Meurger and Gagnon on the lindorm.

1   S.Rosen. *The Dragons of Sweden*. Fate. (date ?)
2   ibid pp.37-38
3   ibid p.38
4   ibid p.40
5   ibid p.42
6   ibid p.43
7   ibid p.45

For reasons too tiresome and unexciting there are no lyrics today, I hope there will be tomorrow!

1 COMMENT:

janne said...
Hi there, happened to read this. I'm trying to collect as many of these kind of stories as possible. My theory regarding lindorms and lake monsters in Sweden is that they relate to very large or gigantic eels! This includes mythological beings like Näcken, Bäckahästen etc.

SATURDAY, DECEMBER 19, 2009

MUIRHEAD'S MYSTERIES: Three quasi-Fortean stories

Today I am presenting these stories in reverse chronological order and you will soon understand why: The first account, in the form of an email from Tony Whitaker a few days ago connected to the attempts by scientists in New Zealand to determine whether or not the giant gecko, or kawekaweau as the Maoris call it, is still alive. (It is encouraging that scientists 'down under' are prepared to talk to cryptozoologists without dismissing us as lunatic fringe.) The following two bits of information I took out of a file at random, then noticed they were both dated around now as far as time of month is concerned!

Firstly, the kawekaweau: 'Hello Richard. There have been no further reports of large lizards anywhere in New Zealand that I'm aware of in the last 18 years. Both Bruce and I, as experienced herpetologists, believe that there were no exceptionally unusual animals around the Tologa Bay area...in our naivety we were trapped in a media and diplomacy pincer-movement.

'The best hope...only hope...to really advance the kawekaweau story will come from either/or:

1. The discovery of fossil or subfossil remains
2. A molecular analysis of the Marseille specimen to determine its true phylogenetic status
3. The discovery of personal letters or diaries in an archive somewhere that provides added information on 18th or 19th century sightings
4. Discovery of documentation around the accession of the sole specimen into Marseille Museum.'[1]

I'm tempted to follow up option 3 myself.

The next item relates to `Dinosaurs rearing young`. I am sure much more is known about this now (note the date was 1991) but it is still interesting.

'Dinosaurs hatched their eggs and cared for their young, according to scientists who

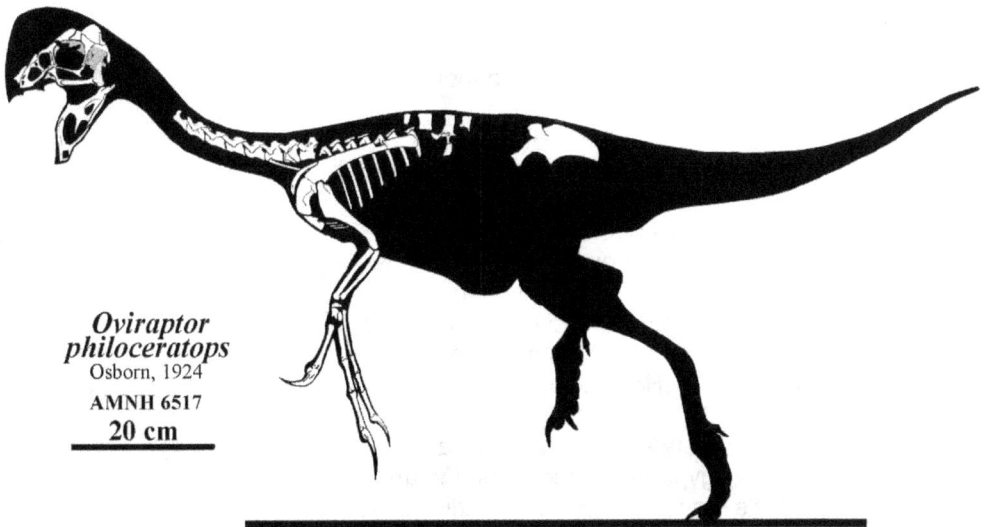

Oviraptor philoceratops
Osborn, 1924
AMNH 6517
20 cm

have discovered the fossilised remains of one sitting on a nest. The fossil of Oviraptor, killed during a sandstorm while hatching its eggs about 75 million years ago, has delighted palaeontlogists worldwide and provided the most graphic evidence so far of the bird-like habits of the dinosaurs. The find has also put to rest the theory that dinosaurs simply laid their eggs in the ground and left their young to fend for themselves...Robin Cocks, Keeper of Palaeontology at the Natural History Museum in London, described the find as "the clincher." "We`ve always suspected this happened but we`ve never seen it. We know about the eggs and we`ve seen dinosaurs close to them but we`ve never seen the two together before. It`s very exciting," he added.' [2]

Finally, remember those Cold War intrusions of mystery under water objects in Swedish waters? Well in December 1991 there emerged (excuse the pun) a possible explanation:

'Submarines, some of unusual design, attacked the Swedish coastal defence system in the Baltic several times in the 1980s, deliberately damaging electronic sensors in the sea bed, Swedish authorities have confirmed. The Swedes have not solved the mystery of the prowling submarines, as they were unable to confirm who was responsible. A special Submarine Commission acknowledged some reports of "alien submarine activity" detected by sonar, were, in fact, swimming minks...In the mid-1980s a mined area off northern Sweden was tampered with and put out of action. The Commission said electric equipment at great depth had been damaged by blows from a hard object. Although some reports were explained by mink, others were clearly caused by man-made submarine activity.' [3]

1. E-mail from Tony Whitaker to Richard Muirhead December 16th 2009
2. D.Penman.Dinosaurs `reared young` *Independent* December 21st 1995.
3. C.Bellamy. Furry clue in mystery of the Baltic Prowlers. *Independent* December 23rd 1995.

Bob Dylan `Neighbourhood Bully`

Well, the Neighbourhood Bully,
He's just one man,
His enemies say he's on their land
They got him outnumbered
A million to one
He got no place to escape to,
no place to run
He's the neighbourhood bully.

Every empire that's enslaved him is gone
Egypt and Rome, even the great Babylon,
He's made a garden of paradise in the desert sand,

In bed with nobody, under no one's command,
He's the neighbourhood bully.

What has he done to wear so many scars?
Does he change the course of rivers?
Does he pollute the moon and stars?
Neighbourhood bully, standing on the hill
running out the clock, time standing still
Neighbourhood bully.

Nighty-night, Richard

1 COMMENT:

shiva said...
"2.A molecular analysis of the Marseille specimen to determine its true phylogenetic status"

Surely this won't determine where the specimen is "from"?

OK, by establishing its phylogenetic relationships with other geckos, it might give a rough idea of whether it's "plausible" for it to be a New Zealand species (i.e., if its closest relatives are from somewhere entirely different, then it's unlikely), but that's very far from proof. Only options 1, 3 or 4 will conclusively place *H. delcourti* as a New Zealand species.

SUNDAY, DECEMBER 20, 2009

MUIRHEAD'S MYSTERIES: Large black birds in Uganda.

Hello, cryptopals! How's it going?

I came across an old-looking print-out of a series of email messages between Ben Roesch and someone who just appears to be called `dwn194`. I have a vague memory that Dr. Darren Naish passed on these details to me but I'm not absolutely sure. So could dwn194 be Darren? Probably!

The first email is dated November 16th 1995:

'This interesting cryptozoological report comes from *Touching the Moon*, by John Preston, pp. 35. the book is about Preston`s interest in travelling to the Mountains of the Moons in Uganda, Africa, and his fulfilment of his dream. I thought some of you would be

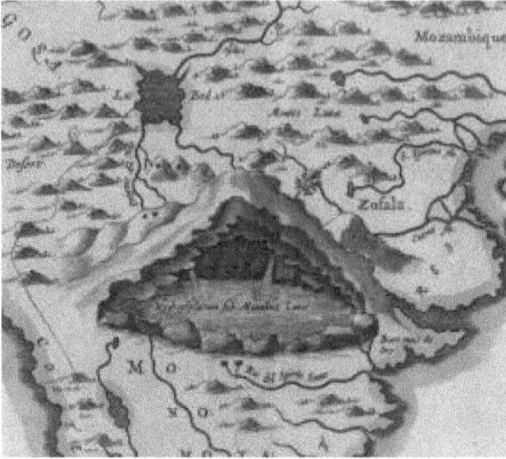

interested in hearing this:

'In 1896 one S. Begge climbed up to a height of 9,000 feet. His servant climbed higher and came back with reports of finding a small lake on the shores of which were a number of black birds the size of sheep. When he tried to get close to them, they bellowed at him like bulls and he ran away. Far from thinking his servant was mad, or effected with altitude sickness, Begge was only sad that he had not seen the black birds himself.'

'First, I might remark that the area of the Mountains of the Moons is the home to plant species that have grown to remarkably large heights and size. Groundseld grow to 20 feet in height and heather "mighty as trees" (in the words of botanist Patrick Millington, who went there in 1934). The area is covered with mists and giant forests of plants that are normally small in the western world. It sounds like a perfect cryptozoologist's (or zoologist or botanist) playground. Who knows what may lie here?

'I must say that a very thorough search of these mountains was performed in 1906 by an explorer named Duke of Abruzzi. He found no evidence of these "black birds" but I don`t know of his other zoological and botanical discoveries he may of made. [1]

1.E-mail from Ben Roesch to dwn194 November 16th 1995.

Talking Heads `Life During Wartime`

Heard of a van that is loaded with weapons
Packed up and ready to go
Heard of some gravesites, out by the highway
A place where nobody knows

1 COMMENT:

Darren Naish said...
'dwn194' is indeed Darren Naish. The report you cite here was covered by Ben in *The Cryptozoology Review*. See...

Roesch, B. S. 1997. A compendium of cryptids (a Russian lake monster; rare shark rediscovered in Borneo; the thylacine – everywhere but in Tasmania; the mamlambo – a "man-eating" reptile?; big black birds in the Ruwenzori;

east coast Caddy?; notes of various new and rediscovered species; assorted news). *The Cryptozoology Review* 2 (1), 4-16.

MONDAY, DECEMBER 21, 2009

MUIRHEAD'S MYSTERIES: A New species of horse in Tibet mid 1990s.

I have a series of E-mails between Ben Roesch and Dr. Darren Naish continuing the selection begun about the large black birds in Uganda. This time I focus on a new species of horse discovered in Tibet:

'An expedition to the far north of Tibet led by Dr Michel Peissel, a Frenchman with more than a touch of Indiana Jones about him, has discovered a hitherto unknown ancient breed of horse. The Riwoche horse, named by Dr Peissel after the remote area where it was found, may be the missing link between the Przewalski horse, a wild Mongolian animal with neolithic origins, and other breeds.

'The team of six, which included Sebastian Guinness, with whom Dr Peissel discovered the source of the Mekong last year, returned to Europe a few days ago. The original purpose of the seven-week expedition was to study another horse, the Nangchen, identified by Dr Peissel in north-eastern Tibet in 1993.

'He had hoped to buy some of theses pure-bred creatures which have no trace of Mongolian, Arab or Turkish blood. Powerful and fast, they have many of the characteristics of a modern racehorse.

'The high prices wanted by the tribesman made purchase impossible, but bad weather on the way back to Lhasa led to a new discovery. "We were in a very unexplored area, the primitive pre-Buddhist area of Tibet to the north of Lhasa, not far from the Chinese border. We weren`t able to proceed on our intended route because the passes were blocked with snow. So we took another way, into Riwoche, which is where we found the little monster. "It is pony-size, about 4 ft, a little like a donkey but with small ears, hardly any nostrils and a rough coat. It has a black stripe down its back, stripes on its back legs and a black mane. I thought it looked like cave drawings of horses, although a friend of mine says it looks like a pig...Dr Peissel, who is fluent in Tibetan, hopes to return to Tibet next year in co-operation with the Chinese Academy of Science to conduct a further study of the horses and to export some.' [1]

1. E-mail from Ben Roesch to Darren Naish and others.Nov.16th 1995.

Muirhead`s Mysteries will be taking a break from December 23rd to December 27th inclusive. The blogs will resume between Dec.28th and Dec.30th. No blog on Dec.31st, Jan.1st and Jan. 2nd. I will resume after the New Year on January 3rd 2010.

Suzanne Vega `Wooden Horse`.

I came out of the darkness
Holding one thing
A small white wooden horse
I'd been holding inside

And when I'm dead
If you could tell them this
That what was wood became alive
What was wood became alive

1 COMMENT:

shiva said...
Apart from the stripes on the legs, it sounds a lot like the Kiang, or Tibetan wild ass: http://en.wikipedia.org/wiki/Kiang

The only otherwise unstriped equid which has stripes on its legs is the Somali wild ass: http://en.wikipedia.org/wiki/File:Somali_Wild_Ass.JPG (which is a subspecies of the ancestor of the domestic donkey).

From the pictures on Wikipedia, the Kiang is by far the most "horse-like" of the wild asses in appearance (to me it looks a lot more like a pony than a donkey). I'd imagine that with a local horse they could produce mules which would look "horsey" enough for their hybrid status not to be obvious, and if they're genetically closer to horses then enough of the mules might be fertile to produce a "horse" breed with some Kiang bloodlines in its ancestry...

TUESDAY, DECEMBER 22, 2009

MUIRHEAD'S MUIRHEAD:Gleanings from Folklore Frontiers no 62

A few days ago I received the current issue of *Folklore Frontiers*, number 62. There are a few interesting Fortean zoological gleanings in it, which I thought worth recording here in this, my final Muirhead's Mysteries before Yuletide. I hope you all have a Merry Christmas and a Happy New Year. The M-Files (stuff the X-Files, what about the M-Files!?) will resume on December 28th for a few days, then properly in the New Year on January 3rd 2010.

SHELL SHOCK: 'A contestant in World Egg Throwing championship, held in Swaton, Lincs., was banned for using a bazooka.' (Daily Star,25/6/09) [1]

BEES: David Hambling, discussing colony collapse disorder, mentions that the disappearance of bees was a running theme during the 2008 season of *Dr. Who*, where they were an alien species (*Fortean Times*, No 255, 2009). It reminded me of a claim about 35 years ago in *The Atlantean,* which stated that bees came from Mars and this accounted for their aerodynamic characteristics being impossible for Earth scientists to explain. Hambling also touched upon the controversial (and unauthenticated) quote attributed to Albert Einstein that if bees died out mankind would survive no more than four years (or was it also a shorter timespan?) [2]

GREYFRIARS BOBBY: (Screeton-Mars Bars and Mushy peas p.113) James Cormack, 78, a chauffeur from Edinburgh recalled how he squired the former filmstar turned animal activist Brigitte Bardot when she visited Scotland ten years ago to help save Woofie, the collier-boxer mongrel ordered by a court to be destroyed after it chased and intimidated a postman. "I took the opportunity to tell her the story of Greyfriars Bobby, which she hadn't heard before. Even though she was hearing it through an interpreter she was in tears," said Mr Cormack. "If the decision had gone against the dog Miss Bardot had a scheme to get it out of the country, flying Woofie to St. Tropez in her private plane."' (*The Sunday Post* 10/8/09) [3]

PARAKEETS (FF 44:12, passim). 'It is now 40 years since ring-necked parakeets began breeding in Britain and speculation as to their origin continues. Nature columnist Derwent May trotted out this hoary chestnut: 'One legend is that the first of them got away from Shepperton Studios when *The African Queen*, with Humphrey Bogart and Katharine Hepburn, was being filmed' (*The Times* Weekend 11/4/09) [4]

This final entry has nothing to do with Fortean Zoology but I thought I'd add it anyway:

MARS BARS AND MUSHY PEAS JOKE: 'A spaceship landed in my garden last night and a small creature emerged covered from head to toe in chocolate. I said: "Are you from Mars?"' [5]

1   *Folklore Frontiers* no.62. p.4

2   Ibid p.5
3   Ibid. p.11
4   Ibid. p.12
5   Ibid. p.11

Happy Christmas to all readers!!

Steeleye Span – `Gaudate`

Gaudate, gaudate Christos est natus
Ex Maria virginae, gaudate
Tempus ad est gratiae hoc quod optabamus,
Carmina laetitiae devote redamus
Gaudate, gaudate Christos est natus

SATURDAY, DECEMBER 26, 2009

The Killarney lake monster by Richard Muirhead

In my room, I gleefully pull the lever hidden
Under my desk, an attempt to change history again, (oh, when will I ever succeed?)
The lever transports me to a copse, by an Irish lake
Hidden from view from humanity.

On my journey there,
In a U.F.O. in mid air
I have changed into black clothes and a red ziggurat hat
to look like Mark Mothesbaugh of Devo.

Suddenly I madly gesticulate
And `Whip It` plays out loud
Stirring the murky depths
And Cuddles emerges, to playfully sport
On the autumn waves,
Whilst several Forteans exult and record the event for posterity.

Several months later,
During the freezing cold winter of 2009-2010
The Natural History Museum, London, is hushed, - in awe of the strange skull
But listen – (noises off)
`Space junk `Space junk`!
Oh God it's Richard singing Devo again!

MONDAY   DECEMBER   28,   2009

MUIRHEAD'S MYSTERIES: Mastodon and Mammoth survival

Goody goody gumdrops,

Muirhead`s Mysteries is back, up to and including December 30th, then a gap of a few days until January 3rd, then onwards and upwards! Today's blog is based upon an e-mail from Andrew Ste Marie, an American cryptozoologist, dated September 23rd 2009, concerning living mammoth and mastodon sightings and hoaxes. I am quoting from as much of his e-mail that is relevant and that I have the mental energy for. I also include website links that are relevant.

> 'You asked for information on mammoth survival and it is my pleasure to send you this list of material I have gathered. Woolly Mammoths *(Mammuthus primigenius)* are generally believed to have gone extinct c. 11,000 years ago in North America,10,000 years ago in Siberia, and 4,000 years ago on Wrangel Island. American Mastodons *(Mammut americanum)* are supposed to have gone extinct c. 10,000 years ago. Here are some of my references from my paper on the recent survival of mammoths and mastodons. Many of the references were to miscellaneous facts about mammoths or the Siberian climate, so I omitted those in this list...Most of these sources are available on-line.' (1)

> Living Mammoth sightings & hoaxes

- Silverberg, Robert, 1970. *Mammoths Mastodons and Man*, McGraw Hill Book Company (this is the best one I`ve come across so far for sightings of living mammoths. It is something of a children`s book but it has serious information, seriously written. Unfortunately, it spends a great deal of time mocking those who believe the Bible. I think he got his information from Heuvelmans's *On The Track of Unknown Animals*. Includes information on the Henry Tukeman living mammoth hoax.

- Lister, Adrian and Paul Bahn, 1994. *Mammoths*, Macmillan Publishing Company (I would rate this one as second-best to Silverberg`s for cryptozoological information)

- Krystek, Lee 1996. *Of Mastodons, Mammoths and Other Giants of the Pleistocene*,www.unmuseum.mus.pa.us/mastodon.htm(accessed June 29,2009) And this one would be third-best, but it also mentions a sighting of a possible living glyptodont.

- Anonymous, September 1993 "Are mammoths still alive ?", http://www.answersingenesis.org/creation/v15/i4/mammoths.asp (Accessed May 26, 2009) (discusses the mammoths of Wrangel Island and a very seldom-reported mammoth sighting – this is the only place I`ve found anything about this particular encounter. Highly recommended article.)

- Anonymous,1893. "Mastodons Still Living," *Winnipeg Daily Free Press*, March 28,1893. www.cryptomundo.com/crypto-news/mastodons-alive/ (Accessed June 19,2009) (really interesting mastodon sighting)

- Anonymous,1897. "Do Mastodons Exist? – Good evidence that at least one specimen still lives," *Decatur Daily Republican*, March 29 1897, http://www.cryptomundo.com/crypto-news/mastodon-surv/(Accessed June 19, 2009)

Also, I have heard that there was a Soviet Air Force sighting of a living mammoth in the 1940s, but I have found no reputable source of information on this sighting. Perhaps Heuvelmans's book discusses it….

Recent Artefacts Showing Mammoths

- Swift, Dennis, March 1997. "Messages on Stone" http://www.answersingenesis.org/creation/v19/i2/stone.asp (Accessed May 26, 2000)
- Swift, Dennis, September 1997." Mammoth among the pharoahs?" www.answersingenesis.org/creation/v19/i4/mammoth.asp(Accessed June 16, 2009) both of these articles by Dennis Swift is highly recommended and very interesting. Also see Dennis Swift's website http://www.dinosandhumans.org/ (2)

1.   E-mail from Andrew Ste Marie to Richard Muirhead (1)
2.   Ibid.

Buggles - `Video Killed The Radio Star`

I heard you on the wireless back in Fifty Two
Lying awake intently tuning in on you
If I was young it didn't stop you coming through

Oh-a-oh

They took the credit for your second symphony
Rewritten by machine and new technology,
And now I understand the problems you can see

Oh-a-oh

3 COMMENTS:

Dale Drinnon said...
There are a few scattered continuing rumors of persisting mammoths and mastodons. The most recent New World C14 dates for persisting Proboscidians are of about the time of Classical Athens for isolated Mexican

mammoths and possibly as late as Roman times in Peru. This has been used to justify certain disputed old artistic representations but Bjorn Kurten discusses the matter in one of his books about Ice-age mammals. There is one site in Peru where a mastodon was killed, butchered, and then cooked in pottery vessels of assumedly Roman date: There are also isolated reports of elephants from South America made by explorers in earlier years. I was just looking at Harold T. Wilkins' book about Secret Cities of Old South America and it mentions sightings (I was not saying it is strong evidence here but merely repeating what evidence I have heard)

The most recent evidence from Alaska persists no later than the WWI era as far as I can tell (footprints), but WWII fliers over Siberia were supposed to have reported observing some. In this case and in some of the explorer's South American sightings, the large creatures were seen at a distance and the identification could be suspicious for that reason. There is however some recent evidence from Western Siberia and even the area bordering Kazakhstan that at least a folk-memory of woolly mammoths persists. Apparantly they are called Leschies or Forest Devils there. They are distinctly depicted as being in the shape of red elephants.

fatchance 2005 said...
This is an item from Native American folklore

http://www.americanfolklore.net/folktales/bc2.html

There are other references

" NIDAWI: Baby name books claim that this name means "fairy" in the Omaha language. According to an Omaha friend, nidawi̧ actually means "elephant woman." In the past, this name probably had a more dignified sense to it--anthropologist Alice Fletcher said it referred to a "mysterious or fabulous being," and Osage scholar Francis LaFlesche wrote that the Osage used the same word, nida (without the feminine ending -wi̧), to refer to giant bones they found in the riverbanks. Despite the higher cachet of that story, I'm still not sure a modern girl would be pleased at being named "mammoth woman" or "giant creature woman." There really are sprite or fairy-like beings in the folktales of the Siouan tribes, but nidawi̧ is not one of them. Whatever real or mythological creature nida originally referred to, it was definitely something known for being enormous. "

http://www.native-languages.org/wrongnames.htm

fatchance 2005 said...
Here is another good one

"...In 1934, Strong published a convincing article detailing the Native

American knowledge of the wooly mammoth. The Naskapi describe a monster they call Kátcheetokúskw (present in many of their myths) as being very large, having a big head, large ears and teeth, and a long nose with which he hit people. When presented with photos of modern elephants, the informants said they fit the description of Kátcheetokúskw as represented in their oral history. The Penobscot of Maine describe a huge animal with long teeth that leaned against certain trees to sleep (noting that when these beasts lay down, they could not get back up). The Ojibwa and Iroquois note the existence of a large beast that once ranged through the forest and was so strong that it would easily knock down any trees that stood in its path. These "elephant" legends are rampant in many other Indigenous cultures such as the Micmac, Alabama, Koasati, and Chitimacha..."

http://nativehistory.tripod.com/id15.html

TUESDAY, DECEMBER 29, 2009

MUIRHEAD'S MYSTERIES: Giant dragonflies

Hello.

Today I feature a brief correspondence that appeared on the *Fortean Times* cryptozoology forum between myself and some others in June and July 2008 on the topic of giant dragonflies. In those days I called myself Dickydoubt_7. I now am known as Dickydevo. I have left the spelling as in the original.

Posted June 25th 2008: 'Hello! I wonder if anyone can help? I have obtained 2 reports of giant dragonflies from the United Kingdom, [one of these was from Oll Lewis; I believe Oll said it was in Wales, if my memory serves me correctly. See post by LividBullseye, below] from friends who I consider to be trustworthy. If anyone has any reports from their localities, please can they let me know on this Forum? Thank you. If I receive enough reports I will write a piece for *Fortean Times*.' [I never did]

gncxx: How big is giant? Like prehistoric, two-foot wingspan giant?
LividBullseye: Ask Oll Lewis he`s seen one. Find him here or on the CFZ.
nyarlathotepsub 2: I`ve seen these http://glzmodo.com/gadgets/gadgets/wowwee-dragonfly-on-sale-now-for-49-235178.php
disgruntledgoth: about the biggest I have seen had a 6 inch wing span
Dickydoubt_7: I should have said that the dragonfly seen in Oxford measured 12 inches from the tip of one wing to the other.
Peripart: I`m genuinely not being facetious or otherwise dismissive, but it wasn`t one of those radiocontrolled toy dragonflies, was it?Some of them are very lifelike (apart from their huge size) Waylander28 12 inches, that is big..

I had a dried carcass of a dragonfly, found on the sands of the lakes Blessington, (Pulaphuca Lakes to be exact) in Wicklow in Ireland. It was larger than my hand at the

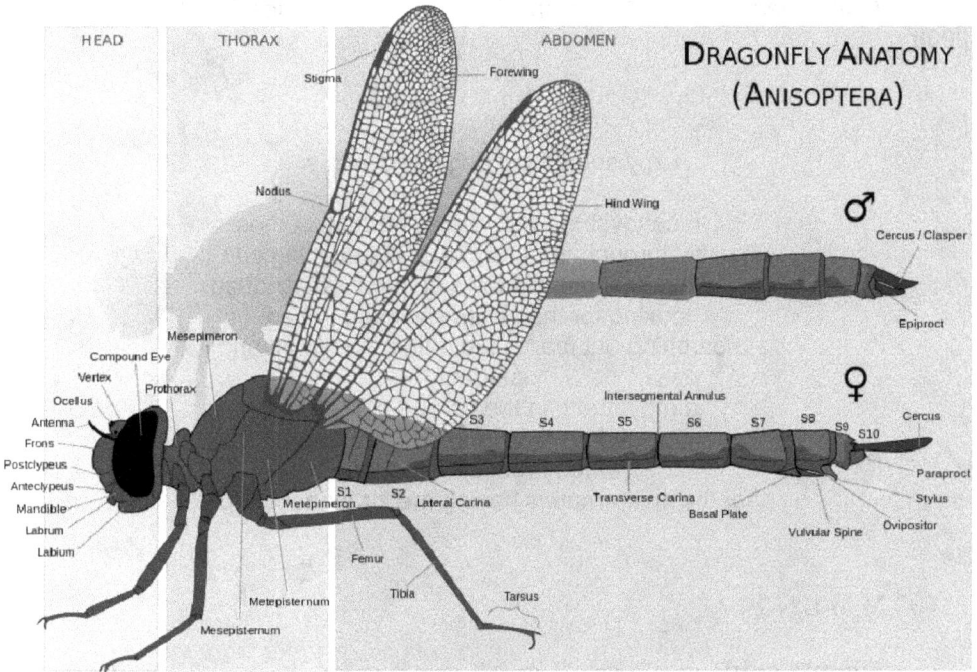

HEAD   THORAX   ABDOMEN   DRAGONFLY ANATOMY (ANISOPTERA)

time, if I can remember it must have been at least 4 and a half to 5 inches long, with a similar wingspan.

I came across it just lying on the sand behind a rock, myself and my friend would not go near it until we were sure it was dead, (we threw a few stones gently around it) it was perfectly poised, full stretch, and wings splayed out to the sides. Long since been reduced to dust now! Shame it was a rare perfect find.

R2800 In the UK? I doubt they could really get that Big honestly. What insects need to become monsters is mostly swamplike humid environments with lots of vegitation and shelter. That way there is no winter to kill them off or hinder their growth, plus the air in swamps seem to have higher Carbon content in the air itself...which seems to produce some pretty big bugs.

300 million years ago there we`re 6 foot wingspan Dragon flies and 6 foot centipedes. And if the world was still as warm as it was back then, they`d still be around.

R2800 wrote: In the UK? I doubt they could really get that Big honestly.

Waylander 28: No it was not in the UK, it was in Ireland. Regardless, they were that big, so much so that from a distance and watching them flying over head, we had mistaken them for sparrows. The body of the Dragonfly that I found was at least half 15mm at its thickest. [1]

1.   *Fortean Times* website discussion. Giant dragonflies. June 25th 2008 to July 29th 2008.

The Cure `Lullaby`

I spy something beginning with s....

On candystripe legs the spiderman comes
softly through the shadow of the evening sun
stealing past the windows of the blissfully dead
looking for the victim shivering in bed
searching out fear in the gathering gloom and
suddenly
a movement in the corner of the room!
and there is nothing I can do
when I realize with fright
that the spiderman is having me for dinner tonight!

1 COMMENT:

Quanta said...
Just wanted to let you know I posted an article at my website yesterday that may interest you. The title of the post is "Giant Dragonfly Encounter in Eastern Kansas" and you can read it here:
http://inter-intelligence-communications.com/?p=1231.

WEDNESDAY, DECEMBER 30, 2009

MUIRHEAD'S MYSTERIES: Exclusive! A hitherto overlooked sea monster off Alaska - 1903

I have recently come across a highly interesting and useful web-based archive of American newspapers http://chroniclingamerica.loc.gov/newspapers/ which has enabled me, in the last few days to find a number of 'new' cryptozoological reports, namely on flying snakes in N. America, which I have passed onto Nick Sucik, probably the world`s leading authority on flying snakes. I also found today`s report on some kind of sea monster, which attacked rorqual whales near Admiraly (i.e. Admiralty) Island near Alaska in the summer of 1903.

I spoke to Dr Darren Naish on the evening of December 28th and he couldn`t identify the creature. He suggested I look at *There are Giants in The Sea* by Michael Bright but it wasn`t mentioned in that book either. Whatever it was, it used a huge 'club' to attack the rorquals.

January ??, 190? THE FALLS CITY TRIBUNE 9

ENEMY OF WHALES.

Strange Creatures Said to Exist in Alaskan Waters.

While operating a fishery on Admiralty island, Alaska, last summer, says a writer, my attention and the attention of my fishing crew was almost daily attracted to a large marine creature that would appear in the main channel of Seymour canal and our immediate vicinity. There are large numbers of whales of the species referred to, and the monster seemed to be their natural enemy. The whales generally travel in schools, and while at the surface to blow one would be singled out and attacked by the fish, and a battle was soon in order.

It is the nature of the rorqual to make three blows at intervals of from two to three minutes each, and then sound deep and stay beneath the surface for 30 or 40 minutes. As a whale would come to the surface, there would appear always at the whale's right side and just above where his head would connect with the body, a great, long tail or fin, "judged by five fishermen and a number of Indians after seeing about 15 times at various distances," to be about 24 feet long, 2½ feet wide at the end, and tapering down to the water, when it seemed to be about 18 inches in diameter, looking very much like the blade of the fan of an old fashioned Dutch windmill.

The great club was used on the back of the unfortunate whale in such a manner that it was a wonder to me that every whale attacked was not instantly killed. Its operator seemed to have perfect control of its movements, and would bend it back till the end would touch the water forming a horseshoe loop, then with a sweep it would be straightened and brought over and down on the back of the whale with a whack that could be heard for several miles. If the whale was fortunate enough to submerge his body before the blows came, the spray would fly to a distance of 100 feet from the effect of the strike, making a report as loud as a yacht's signal gun.

What seemed most remarkable to me was that no matter which way the attacked whale went, or how fast the usual speed is about 14 knots) that great club would follow right along by its side and deliver those tremendous blows at intervals of about four or five seconds. It would always get in from three to five blows at each of the three times the whale would come to the surface to blow. The whale would generally rid itself of the enemy when it took its deep sound, especially if the water was 40 fathoms or more deep. During the day the attack was always off shore, but at night the whales would be attacked in the bay and within 400 yards of the fishery.

"I do not know of any whales being killed, but there were several that had great holes and sores on their backs. Questioning the Indians about it, I was told that there was only one, that it had been there for many years, and that it once attacked an Indian canoe and with one stroke of the great club smashed the canoe into splinters, killing and drowning several of its occupants.

Bridal Shirts.

The Scandinavian bridegroom presents to his betrothed a prayer book and many other gifts. She in turn gives him, especially in Sweden, a shirt, and this he invariably wears on his wedding day. Afterwards he puts it away and in no circumstances would he wear it again while alive. But he wears it in his grave, and there are Swedes who earnestly believe not only in the resurrection of the body, but in the veritable resurrection of the betrothal shirts of such husbands as have never broken their marriage vows. The Swedish widower must destroy on the eve of his second marriage the bridal shirt which his first wife gave him.

An Ounce of Seed.

An ounce of onion seed was sown in the garden of Miss Catherine, at Springfield, Newton Abbot, last March. Recently the gardener gathered 460 pounds of onions.

Sometimes.

Sometimes it is difficult to distinguish between contentment and laziness.

A LACK OF EYEGLASSES.

Returning Traveler Complains of What He Found in British Isles.

The clergyman took off his eyeglasses and carefully wiped them with his silk handkerchief, says the New York Tribune.

"The next time I spend the summer in the United Kingdom of Great Britain and Ireland," he remarked, with vindictiveness in his tones, "I shall carry with me several extra pairs of my particular lenses. Maybe you don't understand what it means to break your glasses in the territory of Edward VII?"

The listener, not being afflicted with eyeglasses, shook his head. The clergyman continued:

"I have been abroad about ten times, and seven of those ten trips were aimed at the British Isles. I have been through Ireland, Scotland, England and Wales, and I have never seen offered for sale in country, hamlet or city store a pair of light weight silver or gold rimmed spectacles, and as for the rimless variety, such as these are, I don't believe the English opticians know what they are, judging from the way in which they stare when an American takes in a pair for fixing.

"The only kind of spectacles worn in the British Isles, as far as my observation goes, is the iron or steel variety, such as our grandfathers and mothers used to hang upon their noses when they perused the Philadelphia and Boston news letters. I repeat that the light weight gold and silver rimmed spectacles are almost unknown in England. You can get them made for you, but you cannot buy them over the counters.

"You ask why? Well, I cannot tell you that any more than I can explain to you why the English insist on riding in the old fashioned apartment carriages on the railroads. I have my own private opinion on the subject, however, and it is this—the cause is to be found in the thriftiness of the people. We are more careless, and a broken pair of spectacles is so ordinary an incident of everyday life here that the man who is forced to wear them generally keeps himself supplied with two or three pairs for emergency's sake. In England the breaking of a pair of spectacles is viewed in the light of a calamity. It is no land for the oculist. The steel rimmed glasses have seen to that. Why, you can take a pair of those heavy weight spectacles and dash them around regardless of any danger of a collapse. That is the reason, experience has taught me, for the lack of the light American variety of glasses. It may be satisfactory for the English, but it is annoying for the American tourist, as I have found to my sorrow. Once in Birmingham I had to wait over three trains to get my rimless glasses repaired, and I had an important engagement in London that day. You ask some other man who's been much in England, with spectacles attached, and see if he does not tell you the same story."

BROOK FARM AMENITIES.

Some of the Humors of the Colony of Notables.

Mr. Lindsay Swift, whose work on Brook Farm is really one of the most thorough monographs ever written in the country, reports a legend that one of the younger members or pupils confessed his passion while helping his sweetheart to wash dishes; and Emerson is the authority for stating that as the men danced in the evening, clothespins sometimes dropped from their pockets. Hawthorne wrote to his sister, not without sarcasm: "The whole fraternity eat together and such a delectable way of life has never been seen on earth since the days of the early Christians. We get up at half-past six, dine at half past twelve and go to bed at nine." An element of moral protest also entered into the actual work of the more serious members, writes Thomas Wentworth Higginson, in Atlantic. Thus Mr. Ripley said to Theodore Parker, of John Dwight, afterwards eminent as a musical critic: "There is your accomplished friend; he would ho-corn all Sunday if I would let him, but all Massachusetts could not make him do it on Monday." Ripley adds that Parker replied: "It is good to know that he wants to hoe corn any day of the week." The question is not how far these details were based on fact or were the fruit of fancy, but the immediate point is that they materially aided in keeping up the spirits of the unbelieving world outside.

Darren suggested it could be mating activity, the club being the male's penis or a giant squid. I used the phrase 'strange creature' when I used the search facility.

ENEMY OF WHALES

Strange Creatures Said to Exist in Alascan [sic] Waters

While operating a fishery on Admiraly [sic] Island, Alaska, last summer, says a writer, my attention and the attention of my fishing crew was almost daily attracted to a large marine creature that would appear in the main channel of Seymour canal and our immediate vicinity. There are large numbers of whales of the species rorqual there, and the monster seemed to be their natural enemy. The whales generally travel in schools, and while at the surface to blow on would be singled out and attacked by the fish, and a battle was soon in order.

It is the nature of the rorqual to make three blows at intervals of from two to three minutes each, and then sound deep and stay beneath the surface for 30 to 40 minutes. As a whale would come to the surface, there would appear always at the whale's right side and just above where his head would connect with the body, a great, long tail or fin, "judged by five fishermen and a number of Indians after seeing about 15 times at various distances," to be about 24 feet long, 2½ feet wide at the end, and tapering down to the water, when it seemed to be about 18 inches in diameter, looking very much like the blade of the fan of an old-fashioned Dutch windmill.

The great club was used on the back of the unfortunate whale in such a manner that it was a wonder to me that every whale attacked was not instantly killed. Its operator seemed to have perfect control of its movements, and would bend it back till the end would touch the water forming a horseshoe loop, then with a sweep it would be straightened and brought over and down on the back of the whale with a whack that could be heard for several miles. If the whale was fortunate enough to submerge his body before the blows came, the spray would fly to a distance of 100 feet from the effect of the strike, making a report as loud as a yacht's signal gun.

What seemed most remarkable to me was that no matter which way the attacked whale went, or how fast (the usual speed is about 14 knots) that great club would follow right along by its side and deliver these tremendous blows at intervals of about four or five seconds. It would always get in from three to five blown at each of the three times the whale would come to the surface to blow. The whale would generally rid itself of the enemy when it took its deep sound, especially if the water was 40 fathoms or more deep. During the day the attack was always off shore, but at night the whales would be attacked in the bay and within 400 yards of the fishery.

"I do not know of any whales being killed, but there were several that had great holes and sores on their backs. Questioning the Indians about it, I was told that there was only one, that it had been there for many years, and that it once attacked an Indian canoe and with one stroke of the great club smashed the canoe into splinters, killing and drowning several of its occupants."

I found this story in the *Leavenworth Echo*, Leavenworth, Washington, and January 30th 1914: Petrified Animal in Mine. A petrified body, apparently that of a seal, was found at a depth of 176 feet in a mine at Carthage, Mo. The head resembles that of a calf, but the body is shaped like a seal. The strange creature is now on exhibition in Carthage. [2]

It would be interesting to see if the age of the 'seal' and the age of the mine strata coincided.

(1)   The *Falls City Tribune*. January 22nd 1904. (Falls City, Nebraska)
(2)   *Leavenworth Echo* January 30th 1914.

Talking Heads -`Life during Wartime`.

Heard of a van that is loaded with weapons
Packed up and ready to go
Heard of some gravesites, out by the highway,
A place where nobody knows
The sound of gunfire, off in the distance,
I'm getting used to it now
Lived in a brownstone, lived in a ghetto,
I've lived all over this town,...

2 C O M M E N T S :

Dale Drinnon said...
Any chance that big fin would belong to a humpback whale? The size could be misjudged as well as the interpretation of the action going on could be.

Max Blake said...
Male humpbacks are known to slap one another with their pectoral flippers. The huge flipper size (5m maximum) gives them their genus name, Megaptera_, meaning giant wing. It is a common rorqual, so it could be the species of rorqual seen by the witness. Alaska is also within the species' range.

SUNDAY, JANUARY 03, 2010

MUIRHEAD'S MYSTERIES: An odd black wolf in Idaho in 1909

Dear folks,

There is nothing odd about wolves in North America with black coats, but a particular wolf shot in the Boise National Forest in early 1909 did cause surprise, according to *The Standard*, Ogden, Utah, April 8th 1909. (This blog continues the series I began a few days ago with my survey of on-line United States provincial newspapers which can be found on http://chroniclingamerica.loc.gov/newspapers/)

The story goes like this:

(The headline is 'Freak Wolf is a Rare Specimen – Officials of Biological Survey Express Opinion – Strange Creature of Enormous Size and Curious Appearance – Not a Hybrid')

'The district forest officers have received confirmation of their belief that Forest Supervisors E. Grandjean`s freak wolf which he recently shot within the Boise national forest and subsequently forwarded to the biological survey at Washignton DC is indeed a rare specimen of quadruped.

'The strange creature was of enormous size for a member of the wolf family, with its back and other portions of its body covered with a heavy growth of black hair, resembling somewhat the coat of a Newfoundland dog, except that it was heavier and coarser.

'The most curious feature in connection with the animal`s appearance was the fact that it was "bob-tailed". Old hunters and trappers in that part of the country, who examined the beast, stated that they had never heard of nor seen anything like it in all their experience. Even the Indians in that region were unfamiliar with the species.

'It will be recalled in connection with the early reports published upon the return of the Lewis and Clark expedition, that Captain Clark mentioned several times the discovery of rude Indian drawings upon rocks and ancient skins of well known animals of the Rocky Mountain region, also a few that were not known to people of that time. Among these was a giant dog-wolf of ferocious appearance and enormous size. A curling shaggy mane was represented as extending down the back of the animal and in one instance it was pictured bearing a young deer in its jaws, illustrative of its size and strength.

'It is possible that the specimen mentioned above is one of the rare descendants of an almost extinct species of wolf which once infested the Rockies terrorizing the Indians and remaining long in their traditions and picture records.

'The officials of the biological survey at Washington DC express their opinion that the animal who's skin and skull was sent by the supervisor of the Boise forest, was not a hybrid and state that the only specimen that resembles it at all closely is one which came from the Priest river forest in northern Idaho. They are anxious to secure further skulls and pelts and offer a good price for same.' [1]

Another website; *Howling For Justice, Blogging for the Grey Wolf, Black Wolves Result of Long Ago Tryst With Dogs* says 'Between 12,000 to 15,000 years ago wolves bred with their close relative, the dog, who passed on to them the black coat color mutation. Black wolves are almost exclusively unique to North America. The black mutation is not present in Europe or Asia, except for a recent Italian hybridized wolf/dog'. [2]

1 *The Standard*, Ogden, Utah, April 8th 1909
2 Howling for Justice web-site. Downloaded January 2nd 2010.

Steeleye Span 'Female Drummer'

I was brought up in Yorkshire and when I was sixteen
I walked all the way to London and a soldier I became
With me fine cap and feathers, likewise me rattling drum
They learned me to play upon the ra-ba-da-ba-dum
With me gentle waist so slender, me fingers long and small
I could play upon the ra-ba-dum the best of them all

And so many were the pranks that I saw upon the breech
And so boldly did I fight me boys although I'm but a wench
And they buttoned them up me trousers so up to them I smiled
To think I'd live with a thousand men and a maiden all the while

7 COMMENTS:

Dale Drinnon said...
This is interesting, are there any further reports of the Bobtailed wolf? It sounds like the sort of thing that would inspire ghost stories told by campers around the campfire.

I had separately heard legends of a gigantic lynx that used to prey on bison and which is supposed to be depicted in rock art as well, but it was supposed to have been a great cat and not a dog: furthermore despite its characterization as a "Lynx" the part about the tail being bobbed off does not seem to be exactly specified by my information.

It seems to have been called "Buffalo cat", or rather that is the English equivalent.

Marcy said...
This wolf sounds like a shunka warakin.

Retrieverman said...
Some observations:

Wolves vary in appearance as much as dogs do. This could be a very unusual wolf.

The distinctions of dog and wolf are not clear-cut as one might assume. There are doggish wolves and wolfish dogs.

I have several true life accounts of pet wolves that were as easy to handle as the average Labrador, and my grandfather had a dog -- half Norwegian elkhound and toy collie (not a Sheltie) -- that might as well have been a wolf. Unlike either of his parents he didn't bark. He was tolerant of other dogs until they challenged him, and then he was a fierce. He could be petted but if you stepped on his toes by accident, he would bite. (Normal domestic dogs put up with this.) I've seen photos of him. He looks like a coyote with a white ring around his neck. He was not a coydog, for there were no coyotes in my part of the country at the time.

The skull and skin examinations could only tell you the preponderance of its heritage. The wonderful discovery that black wolves and coyotes got their coloration through hybridization with domestic dogs shows these distinctions are rather muddled.

It's also worth checking out the story of the last wolf killed in Scotland. Apparently this was an unusual black wolf with a penchant for attacking children -- a well-known wolf hybrid behaviour. A huntsman named MacQueen of Pall à Chrocain. MacQueen's long dog (probably a cross between a greyhound and deerhound, long dogs are crosses between sight hounds) caught the wolf or wolfdog and MacQueen cut its throat. That was the last wolf of Scotland.

I classify dogs as *Canis lupus familiaris*, just as the domestic cattle of European ancestry are classified as *Bos primigenius taurus*, conspecific with the Aurochsen. The only dogs that have been proven to have golden jackal and coyote in them are those that have actually been bred in captivity, like the Sulimov dog and the some of the so-called Spirit dogs from the American West that have been crossed with coyotes.

shiva said...
How did they know the wolf wasn't a hybrid? They just state that without evidence. It sounds like a hybrid with its dog genes coming from a very large breed of dog (e.g. the Newfoundland (?) from Jack London's "Call of the Wild") to me. The tail could easily have been lost in an accident or a previous failed attempt to shoot or trap it.

The "gigantic lynx that used to prey on bison" sounds a bit like a late-surviving Smilodon - weren't they fairly short-tailed and more lynx-like than lion- or panther-like in limb/body proportions?

Retrieverman said...
Dale,

One of the names of the Canada lynx in the early days of colonization was called the "loup cervier." The name is still used in Francophone Canada.

The name means "deer-like wolf."

Anglophone Canadian thought that they were referring to the term "lucifer," and it's still a common name for the animal, which has tufts on its ear and kind of looks a bit like the devil.

It would not be surprising that a large black wolf with very coarse hair would have its tail removed. When wolves "war" with each other over territory, one of their common tactics of combat is to grab their adversary by its tail. This is such a common tactic, that virtually all livestock guardian dogs in wolf territory are docked. A wolf can grab a dog by the tail if the dog has no tail.

I think what we have here is a very unusual wolf-- perhaps with some distant dog heritage.

Retrieverman said...
The cryptid "wolf" that has captured my imagination is none of these animals. The Andean wolf enigma has still not been answered. It could have been a sheepdog. It could have been a pet or zoo wolf that went wild or was released into the mountains. It could have been a descendant of a vestigial population Dire Wolf. However, it was black, and my guess is that Dire wolves were not black in color.

Retrieverman said...
Found something interesting:

Late 1700's, Western Hudson Bay – "the Dog... resembles the wolf, but in size is greatly inferior... They run and bite in silence, never barking but sometimes howl egregiously... It is usual for our [Newfoundland] dogs and also the native breed to copulate with wolves, and the offspring retain the moroseness of the latter."

Williams, G. ed., intro by R. Glover. 1969:33. *Andrew Graham's Observations on Hudson's Bay*, 1767-91. The Hudson's Bay Record Society, Vol. XXVII, London

http://mainehuntingtoday.com/bbb/2009/08/19/is-there-really-any-such-thing-as-pure-wolf/

European traders, explorers, and trappers almost always had Newfoundland dogs. Newfoundland's varied a lot in appearance, but very few were the great big mastiffs that currently make up the breed. That is mostly a creation of dog dealers in the nineteenth century. The original Newfoundlands varied from the short-haired dog that looked like a Labrador to a larger black and white dog that could haul. These dogs could haul, hunt, and retrieve shot game. All dogs we call retrievers are derived from "Newfoundlands." (The smaller short-haired ones are extinct-- or rather disappeared into the Labrador retriever. We now call the breed the St. John's water dog.)

Shiva,
Buck, the dog in Call of the Wild was a cross between a St. Bernard and a "Scotch Shepherd"-- what we call the rough collie today.

TUESDAY, JANUARY 05, 2010

MUIRHEAD'S MYSTERIES: A Texas wildman and giant lizards

Today I continue with my look at American crypto-strangeness with a report from San Jacino Bottom, which I presume is in Texas, and also the Santa Ana River in California.

The first report is from *The Houston Daily Post* for August 13th 1901, which I reproduce here verbatim:

'Strange Adventure that Befell Commissioner Bucher in San Jacino Bottom. Commissioner J. C. Butcher, who came in from Harrisburg yesterday to attend a meeting of the commissioners` court, was engaged during the recess hours yesterday in relating a strange adventure which befell Mr C. L. Bering and himself a day or so ago while they were hunting in San Jacino bottom. In their search for game they penetrated deeper and deeper into the recesses of the wood and in a spot where the foliage and the long grass were so thick as to almost block their passage, they came face to face with a wild man.

'The strange creature of the wood had long mane like hair falling far below his shoulders and he shook a mat of it from in front of his eyes and glared at the hunters as if intent on disputing their right to the wilds.

'The wild man was clad principally in a yard of sunshine, which struggled through a rift in the trees, and a devilish smile of defiance. Additional to these the strange creature was attired in a breech clout of leaves.

'The hunters were so taken aback by the unexpected presence and the uncouth appearance of the stranger that before they had time to test the conversational powers of the wild man he glided away into the dense forestry without more ado'. [1]

Now from *The Los Angeles Herald* of January 1st 1909:

'Tramps Frightened By Big Alligators. Monster Saurians Appear Among Willows on Santa Ana River and Wayfarers' Camp is Hurriedly Evacuated.

'Either a bunch of fifteen tramps had an extra supply of barleycorn on tap last night, or the willows along the Santa Ana river this side of Riverside are peopled with strange creatures on the monster order. At 2 o'clock this morning a number of hobos arrived here out of breath and thoroughly alarmed, having run most of the distance between Riverside and this city *[i.e. San Bernardino - R]*. They tell of having suddenly been attacked by two immense creatures resembling in general appearance lizards. They were about a campfire at the time, when one of their numbers suddenly emitted a frightened yell and bounded through the fire, disappearing in the darkness. His companions looked to where he had been seated, and, catching a glimpse of the strange creatures, they too fled. The report was investigated this morning and disclosed that in all probability the creatures mentioned are two big alligators which escaped from a Riverside man recently. Word from Riverside states that the alligators have not been seen since, though a systematic search has been instituted.' (2)

These two stories were found on the website I mentioned a day or two ago.

1   *The Houston Daily Post* August 13th 1901
2   *The Los Angeles Herald* January 1st 1909

All for now, Richy

> Steeleye Span 'Hard Times of Old England'
> Come all brother tradesmen that travel along
> O pray, come and tell me where the trade is all gone
> Long time have I travelled, and I cannot find none
> And sing all the hard times of old England
> In old England, very hard times

1 COMMENT:

Dale Drinnon said...

The Wildman account sounds as if it was meant to be one of those "Crazy Indian" reports-he had long hair on his head and a loincloth, but no body hair is mentioned (I would be just as happy if it were a Neanderthal myself); And it would be really useful to know what the tramps were calling "big lizards" (? Iguanas) or how big they were, since the identification as alligators is only an assumption and there ARE otherwise reports of big lizards (that run on their hind legs) in the area.

SUNDAY, JANUARY 10, 2010

MUIRHEAD'S MYSTERIES: The Hodag and an out of place turtle

Dear folks,

Muirhead's Mysteries continues today with the tale of the Hodag and a turtle (or should that be tortoise?) turning up somewhere in the Nevada desert. The text of the newspaper report on the turtle actually used that word, but this was no sea creature as far as I know. However, the hodag was almost certainly mythical if the author of the Wikipedia entry is to be believed, though the newspaper item on this has a more literal stance. In chronological order, the first item is from *The St. Paul Globe* April 20th 1903. I do not quote in full; just the most interesting parts:

'MR OPSHAL SEES QUEER ANIMAL. It haunts the woods about his country place at the lake.

What is a hodag A. H. Opshal asserts excitedly that it is a surviving representative of a supposedly extinct sarian that looks like a cross between an iguana and a rocking horse, and offers to produce witnesses to the strange nocturnal proclivities of the uncanny brute that haunts the woods about his country place, Ruritania, Lake Minnetonka...A lot of lumbermen over in Michigan faked up a hideous looking reptile and had a picture taken of him, which the Northwestern Lumberman printed, but somebody blew the game and the hodag, hideous as an inhabitant of Dante's dread picture, proved to be a stuffed nothing in particular.'

The article then goes on to insist that the hodag is a real animal: 'Opshl himself emptied a double head of B. B shot at the beast or reptile, the range being 30 yards in the moonlight. The shot was heard to rattle like hail on the scaly coat of the creature which promptly emitted a sound which resembled, roughly speaking, a cross between the laugh of a hyena and the bawl of an indignant cow.

"I don't know what breed this creature belongs to" said Mr Opshal yesterday "but what I do know is that there is nothing in the books describing him. He is scally all over like a big fish...He is not in the least injured by being shot at by any sort of small arm...When he runs his tail is held high above his back and it has spines sticking out all along its length like an iguana...The prevailing opinion is that the creature which has created so much discussion at the Point is an escaped specimen from some circus or sideshow. No report of such an escape is remembered, but the people who witnessed the latest "hodag" refuse to be laughed out of countenance. Mr Opshal takes a walk around his place every night in an effort to get another shot at the animal. He has provided himself with a 45-90 Winchester and hopes to report results within a few days' [1]

Wikipedia has the following about the Hodag:

> 'In 1893 newspapers reported the discovery of Hodag in Rhinelander, Wisconsin. It had "the head of a frog, the grinning face of a giant elephant, thick short legs set off by huge claws, the back of a dinosaur, and a long tail with spears at the end."The reports were instigated by well-known Wisconsin timber cruiser and prankster Eugene Shephard, who rounded up a group of local people to capture the animal....' [2]

The next item is about a turtle (sic) found in the Nevada desert, later going missing in California.

I think there is a possible link between this Nevada tortoise and some large tortoises reported in California in the 1930s.

Read on!

> *The San Francisco Call* December 1st 1904: 'TURTLE TAKES ITS DEPARTURE`. Strange creature missing from home of its owner in the University town.

> "Japhet in Search of a Father" has a rival in Berkeley in the frantic search which Arthur I Street is making for his giant turtle, a creature which scientists at the State University have declared is 200 years old and entitled to a place in museum annals as a valuable curio. The animal has disappeared and cannot be found. The turtle was captured in the midst of the Nevada desert several months ago by Street while he (Street) was making the journey overland between Los Angeles and Salt Lake City...He was brought to Berkeley and given quarters at the residence of Mr Street's father, 1517 Shattuck Ave. Professor W. E. Ritter of the department of zoology examined the strange find made by Street and pronounced it to be more than 200 years old an a rare specimen of an almost extinct species of animal life. The professor could give no explanation of how the turtle came to be living in the midst of desolation on a great desert, but it is supposed the palm roots or some other hidden vegetation contributed to its support.

> Mr Street made a pet of his find, quartering it at his father's home on Shattuck avenue. Last night the turtle disappeared and to-day Street has searched Berkeley through trying to get a trace of the vagrant creature, but without success.' [3]

1. *The St.Paul Globe.*April 20th 1903.
2. Anon.Wikipedia.Hodag.http://en.wikipedia.org/wiki/Hodag. (Accessed Jan.8th 2010)
3. *The San Francisco Call.* December 1st 1904

Sorry about the lack of lyrics today. I'll try and remember them on Tuesday.
Rich

TUESDAY, JANUARY 12, 2010

MUIRHEAD'S MYSTERIES: Equatorial West Africa's Ikanda

Dear folks,

Whilst browsing through the Chronicling America website again today, I quickly came across a possible 'new' cryptid, but bear in mind I have only done a cursory Google search for it. In 1896 a zoologist in an equatorial region of West Africa came across what he described as a bear-ape, or Ikanda. The story was published in the *Marietta Daily Leader* of August 17th 1896. The zoologist thought it was a new species of potto or ape-like creature. The description of its fur makes it appear to be a golden potto (*Arctocebus aureus*) but I know nothing about these animals so it could be something else altogether. The zoologist R. L. Garner already knew of two pottos or angwantibos; he thought it was neither but similar to them. Fortunately we have illustrations of parts of its body, namely the hands and feet.

I quote:

'THE IKANDA OR BEAR-APE Prof.Garner Tells of the Queer Little Beast Discovered In Africa. It Never "Loses Its Grip – Strange Story Told By The Natives of How It Is Used for Catching Leopards.

In the forests of Central Africa there live many strange creatures, some of them as yet unknown to science. One of the most singular of these belongs to the Simian family, but is very low down in the scale. I have a specimen which I kept in captivity for a number of months, during which time I gave considerable attention to its habits. So far as I am able to learn, it is an entirely new species of a small group known as slow lemurs. Up to this time there have only been two species of this genus known in Africa, but the specimen in my possession is not identical with either. I have never found a specimen of it in any museum, either in this country or abroad, and I have found no zoologist who has ever seen or known one. It is neither a potto nor an awantibo though it doubtless belongs to that group, being included in the genus of lemuroids called "Arctocebus" ie bear apes.

The native name of this animal is "Ikanda" and it appears to be confined to a small scope of country along the equator in West Central Africa...It is uncertain how far it extends to the interior...In his general appearance the Ikanda is like a

189

miniature bear. It is from this fact that the name "arctocebus" has been conferred upon him. Every movement is exactly like that of a bear, the ears nose and tail are also like those of a bear...The most singular physical features are the hands and feet, which are shown in the drawings. The hand is a perfect human hand in every respect, except the want of an index finger....[the article then goes into a highly detailed account of the hands and feet, which I do not have time to include here - R]

'The body is covered with a dense growth of soft hair, almost like fur. It is a dark brown in color, but somewhat lighter on the under side of the body. The nose is quit bare and very black.
I have never seen one of these animals in the adult, but apparently they grow to be 14 or 15 inches in length [1]... The article then goes on to describe the resting and eating habits of the Ikanda. But there is nothing about its ability to catch leopards!

1.   *Marietta Daily Leader* August 17th 1896.

U2   *'New Year's Day'*

All is quiet on New Year's Day
A world in white is on its way
And I want to be with you
Be with you night and day
Nothing changes on New Year's Day

THURSDAY, JANUARY 14, 2010

MUIRHEAD'S MYSTERIES: Tales of cats

Folks - I was going to continue today with the series of crypto reports from chronicling America but I have found something rather significant on the site this afternoon on which I want to carry out some more research, so I am returning to a theme I brought to your attention a few months ago, curious cats.

I have two stories. Firstly, from the Jackdaw column in *The Guardian* of February 19th 1997: '#'A cat's tale.' (No, I'm not being very original today am I?!)

"The large number of tail–less cats in the Flensburg * area on the German-Danish border is thought to be the result of a bomber crash in 1942. On the night of October 12 that year, 27 Halifaxes of 4 Group attacked the U-boat base at Flensburg, 12 being shot down by a new type of Swiss manufactured Oerlikon flak gun moved into the area a few days earlier.

Kurt Peuschel, then a boy of 14, and some of his school friends, was allowed by the German guard to inspect one of the crashed bombers, believed to be W7717 of 10 Squadron from Melbourne. The guard told them that five of the Canadian crew had been taken prisoner and that two bodies had been removed. The prisoners had asked the guard to look for their cat mascot, a tom, which they would recognise because it had no tail. Kurt and his friends were enlisted in what proved to be a fruitless search for the cat.

Now married and living in Switzerland, Kurt visited his 90-year old mother last year and while with her saw a local TV station report on the large number of cats without tails in the Flensburg area.

The report attributed them to the missing mascot, which was thought to have been obtained when the crew were at a rest centre on the Isle of Man. Taken from the *Air Mile*, the Journal of the RAF Association. Thanks to AJ Lne for spotting this jewel. [1]

(* I had a German girlfriend from Flensburg that I met in Derry. She was deeply into the poetry of John Cooper Clarke. I bet you never knew that!!)

Next, a favourite of mine, old moggies:

" The cat`s whiskers Spike is oldest moggie at 29. A 29 year old ginger and white tomcat called Spike was yesterday crowned Britain`s oldest living moggie. The 10lb puss – who is the equivalent of 203 in human years-, won an entry in the Guinness Book of Records. Owner Mo Elkington claimed Spike ....lived so long because she feeds him trendy "healing" plant Aloe Vera. She said "I put some in his food every day. It keeps his fur healthy and protects him against rheumatism." Aromatherapist Mo,47,of Bridport, Dorset, bought Spike as a kitten for half – a – crown at a market. She only discovered Spike was a record breaker when she took him to a vet. She said: " I`d no idea his age was that unusual but the vet was staggered so I called the record people."

Mo added: "He must be lucky because he was bitten by a huge dog at 19. Vets didn`t think he`d live. Britain`s oldest cat died in Devon in 1957, aged 34. [2]

Of course this is 11 years ago, so by now there may have been a new record.

1. *The Guardian* February 19th 1997.
2. *The Sun.* October 15th 1999.

The Beatles `Blackbird`

Blackbird singing in the dead of night
Take these broken wings and learn to fly
All your life
You were only waiting to be free....

MUIRHEAD'S MYSTERIES: Termites, a bird and more on the 9 legged fish.

Dear folks,

Today I have a mixed bag of items, namely a report on foreign termites near the home of the Dear Leader, an early Takahe report from New Zealand and an opinion concerning a 9-legged fish from Nevada. This opinion comes from Chad Arment in his book *Cryptozoology and The Investigation of Lesser-Known Mystery Animals* pp 177-178.

Now the termites: "Alien termites set up home in Devon. Termites, one of the world's most destructive insect pets, have established what is thought to be the first colony in Britain in a house in rural Devon. The termite, *Reticulitermes lucifugus*, is normally found in southern Europe, southern Russia and North Africa."

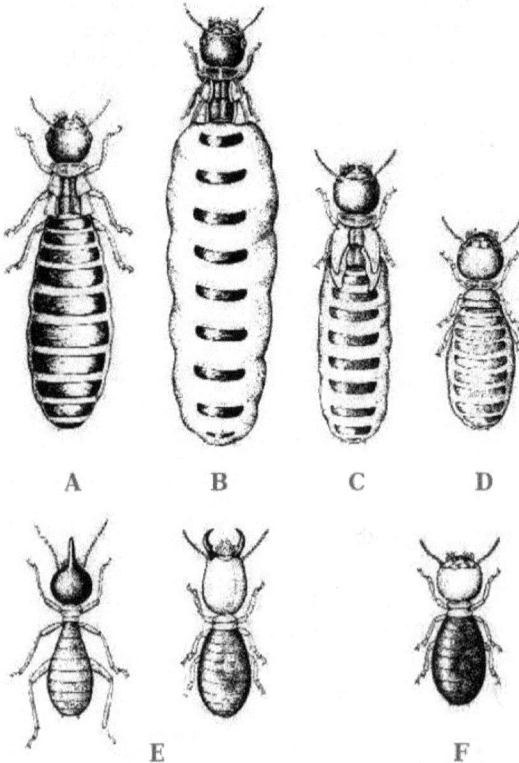

"The termites were found in two timber frame walls, heated by pipes," says Roger Berry of the Building Research Establishment. " The timber was damp and the combination of heat and moisture has produced the perfect microclimate for the insects."

No one knows when the insects set up home in the house near Barnstaple, or how they got there...A colony can contain as many as a million termites. The

Caste system of termites **A** — King **B** — Queen **C** — Secondary queen **D** — Tertiary queen **E** — Soldiers **F** — Worker number in the Devon colony is not known but the investigators counted at least 500 individuals in "just one foot of skirting board." So far no termites have appeared in other properties. "If this were an urban area or a row of terraced houses, we could be looking at a rather more serious problem." says Berry...One thing the experts are sure of is that the termites have not appeared as a result of climate change. "This isn`t a case of change. " This isn`t a case of global warming," says Paul Eggleton, who runs the termite research team at the Natural History Museum. "They have certainly been brought in by somebody." [1]

The Takahe report is from The *San Francisco Call* of October 9th 1898. It's not significant from a cryptozoological perspective, rather, historical. According to Wikipedia the takahe was thought to be extinct after the last 4 were taken in 1898. This article reports on one of those four:

RARE BIRD CAUGHT BY A TOURIST`S DOG. Capture in New Zealand of a Wild Fowl of a Species Thought to Have Been Extinct. Advices were received from Auckland by the Warrimbo to the effect that a species of a very rare bird has been discovered on one of the southwest coast fjords of New Zealand. It is the mantelll, the takahe of the Maoris, supposed to have been extinct. Until a few weeks ago there were only three specimens of this bird in the world, and for many years no trace of others had been seen. The specimen just secured will probably be worth its weight in gold and perhaps a good deal more. It was caught by a dog owned by a tourist.

Information hitherto unknown has been acquired through the capture of this strange creature. It is a handsome bird of heavy gait, and absolutely unable to use its wings for flying. It is about the size of a goose, and has beautifully colored feathers. On the back, wings and tail they are olive green of almost metallic luster and below the short tailfeathers are white. The legs and toes are of a rich salmon red. Another remarkable feature is the beak – a great equilateral triangle of hard pink horn, with one angle directed forward, while on the upper side of the base of the beak is a rudimentary comb. [2]

Now Arment's take on the 9-legged fish. His source was the *Syracuse Post Standard* of August 31st 1905. My source may have been different; the story seems to have been syndicated. Arment says – "Again,* an animal more fabulous than realistic. However in this case, the witnesses truly existed. Sweeney was Attorney General for the state of Nevada during the 1906 San Francisco, California, earthquake. Unfortunately, no mention is made of where the creature is caught. In this case, I'm leaning more toward misidentification than hoax, though I cannot supply a reasonable biological candidate. Perhaps some intrepid investigator will have opportunity to make an appointment with the Smithsonian and try to track the critter down. [3]

* Arment was writing this in 'Miscellaneous Menagerie', a chapter on cryptozoological odds and ends.

1   Alien termites set up home in Devon. *New Scientist* 12th November 1994

2   Rare Bird Caught....*The San Francisco Call*. 9th October 1898
3   C. Arment. *Cryptozoology and The Investigation of Lesser Known Mystery Animals.* (2006) p. 178

I bumped into a young cove in Macclesfield the other day who insisted that Nelson Mandela had been released. Because he is still in jail, I present to you - *The Special AKA* : (*)`Free Nelson Mandela`

Free Nelson Mandela
Free free
Free free free Nelson Mandela
Free Nelson Mandela

Twenty-one years in captivity
His shoes too small to fit his feet
His body abused but his mind is still free
Are you so blind that you cannot see I say
Free Nelson Mandela

[*I can't recall if I was being ignorant here or just plain silly because of course Nelson Mandela was released on February 11th 1990.]

TUESDAY, JANUARY 19, 2010

RICHARD MUIRHEAD: MUIRHEAD'S MYSTERIES: Mystery wild cats, both Irish and Scottish

In August 2008 *Fortean Times* published an article by Gary Cunningham on the subject of the elusive Irish Wild Cat and by this I do not mean escaped pumas or lynxes, but a bona fide Irish Wild Cat. So today, with a little extra money than usual, I paid 10 euros for a 24 hour session on the Irish Newspaper Archive in a step of faith, hoping I would find some references within the many Irish newspapers dating from the 18th century that are archived. I started my search at about 2.30pm today Monday and ended about 10 minutes ago at 8.05 pm and I have made some quite exciting discoveries covering Irish and Scottish wildcats, and Irish snakes, which I will mention here for the first time. (The cats, I mean; not the snakes). The latter are being passed on to Irish cryptozoologist, Ronan Coghlan. There is an interesting comment in the Irish paper *The Nation's* May 22nd 1897 edition: 'Contrary to the popular belief, there are some snakes in Ireland, but they are very rare.' [1]

Just go into Google, type in Irish Newspaper Archive and follow the instructions.

Freemans Journal August 24th 1838

'Mr Cahitl (? Hard to read) of Whiskey-hail, county Limerick, shot, on Thursday last, at Cragg-wood, three wild cats of monstrous size. These strange animals attacked the wood ranger a few days ago, who narrowly escaped with his life. So disfigured were his features, and so completely exhausted was he, that his family did not for a considerable time recognise him on being brought home. Mr Cahitl has taken the skins to send to the Royal Cork Institute.' [2]

I wonder what happened to the Royal Cork Institute?

Southern Star February 11th 1893

'This extract was a summary of a lecture on Irish natural history by the Rev. H Burton Deane at Clonakilty in Co. Cork. In it he says : "Rosscarberry is a capital place for otters, foxes, stoats, weasels, wild cat, and others of the furry tribe."' [3]

Rosscarberry is in Co. Cork.

So here we have wild cats being mentioned in the same breath as well known native Irish mammals.

Southern Star July 19th 1902

[Here we have an interesting corroboration with Cunningham's reports from *Fortean Times*. Cunningham uncovered accounts from Ireland of wild cats with nail-like projections from their tails. For example near Kenry in Limerick (4) and one from Co. Mayo c. 1940-1950 [5]]

But the *Southern Star* report:

> 'Here is a wonderful exemplification of the queer effects which sunstroke has on some timorous minded people: - "Dear Sir, permit me, through the medium of your Macroom Notes to put before the public of Macroon and district a danger which menaces thousands of human lives. There is at present in Coolcower wood a wild cat of the most bloodthirsty description which I saw with my own two eyes on Thursday evening last. This may sound like a hoax, but I pledge you my honour I am perfectly in earnest. I knew the animal as soon as I caught a glimpse of him, crouching as I believed, for a spring upon me. It was at least four feet long with great glaring eyes and a bushy tail about two feet in length with a nail in the end of it. I did not wait for further observations at the time but I can assure you it was a wild cat" Yours truly Pro Bono Publico.' [6]

There then follow a series of facetious comments by the editor, presumably. So here we have the 2nd reference to a Co. Cork wild cat.

The final report for today is from the *Irish Independent* of October 17th 1907, which has a head line – CAT NEARLY 5FT LONG 'A wild cat measuring 4ft 7in has been shot by Mr S A Bland a noted Arizona sportsman.' [7] It does not say whether or not it was shot in Ireland or England.

There is an item in the *Irish Independent* of May 21st 1906 on a wild cat being brought from the Cocos Island. I don't know if this is significant.

The Scottish wild cat. This is from the Anglo-Celt of January 10th 1856. It is interesting because of its colouration:

> 'CAPTURE OF A WILD CAT`. - On Friday week, a man caught, in a wood on Kirkennan-hill, parish of Buittle, a fine live specimen of that nearly extinct class of the savage creatures of Scotland - a wild cat. It had been driven by hunger and the inclemency of the weather from its native retreats into a baited trap. It is of a bluish gray colour, stands high, and measures three feet in length, from the nose to the tip of its tail. – Scotchmen.' [8]

Is this a normal colour for a Scottish wild cat? Seems a bit like a Kellas cat.

I have got up to about 1913 with the wild cat search and will try and bring more to you on Thursday Jan. 21st.

1 *Nation.* May 22nd 1897
2 *Freemans Journal* August 24th 1838
3 *Southern Star* February 11th 1893
4 G.Cunningham. The Irish Wildcat *Fortean Times* August 2008 p.41
5 Ibid p.42
6 *Southern Star* July 19th 1902
7 *Irish Independent* October 17th 1907
8 *Anglo-Celt* January 10th 1856

Now Music time! Simple Minds - `Sanctify Yourself`

Is this the age of the thunder and rage
Can you feel the ground move `round your feet
If you take one step closer, it'll lead to another
The crossroad above is where we meet
I shout out for shelter, I need you for something
The whole world is out there, all on the street
Control yourself, love is all you need
Control yourself, in your eyes
Sanctify yourself, sanctify
Be apart of me, sanctify
Sanctify yourself, sanctify
Sanctify yourself, set yourself free

1 COMMENT:

Kithra said...
"I wonder what happened to the Royal Cork Institute."

If Wikipedia can be believed in this instance the:

"Royal Cork Institution was a Irish cultural institution in the city of Cork from 1803-1885. It consisted of a library of scientific works, a museum with old Irish manuscripts and stones with ogham inscriptions, and lecture and reading rooms. A lack of funds resulted in its closure in 1885."

http://en.wikipedia.org/wiki/Royal_Cork_Institution

So I bet those skins are now long gone.

MUIRHEAD'S MYSTERIES: More on the Irish wild cat and also the Dobhar Chu

I have a bit more information on the Irish wild cat, which in my opinion is a little way closer to being vindicated as once (and still?) existing. Also, the Dobhar chu, or master otter, a giant otter-like animal said to have existed in north-west Ireland and Scotland as recently as the 1700s. It was said to have attacked and killed Grainne–ai–Chonalai who was killed on the shores of Glenade Loch on September 24th 1772. Dr Karl Shuker has written extensively about it, but I have just unearthed a report from the 1950s.

But first the Irish wild cat: I mention an item in the *Irish Naturalist* which was not mentioned by Shuker in *Mystery Cats of The World* or Cunningham in his article 'The Irish Wild Cat' in *Fortean Times* August 2008..

Cunningham mentions the 1906 paper by Dr RF Scharff "On the former occurrence of the African Wildcat (*Felis ochrreata* Gmel) in Ireland, *The Irish Naturalists Journal*." Shuker mentions papers in the same journal from 1905. There was a debate as to whether or not remains of a wild cat found in a cave in County Clare were similar to the African wild cat.

The extract is from the *Irish Naturalist* *for July 1908, vol.17 no 7:

"Supposed Occurrence of a Wild Cat in the West of Cork."

A species of Wild Cat is proved by its fossil remains to have inhabited Ireland at no very remote period, as Dr. Scharff has shown in his very careful paper (*Proc.R.I.Academy*, January 1906), and he also urged that enquiries should be made as to whether such an animal has been seen or heard of lately (*Irish Naturalist*, 1905, p.79). Though the specimen referred to below has unfortunately perished, and conclusive proof of its species is therefore unattainable, it may be well to record the remarkable descriptions given me by several members of the Becher family.

In 1881 I made a note of the statement of Mr E.W.Becher and his sister to the effect that some years previously their elder brother shot a Wild Cat at Liss Ard,the O'Donovan's place "It had a broad head, short legs, bristly tail; the colour "was brindled, with bars of black on a dark grey, with a dash of tan colour"

I have recently met their elder brother, the Rev H.Belcher,who at my request has written the following account:- " Castlehaven Rectory, Skibbereen,May 8th,1898.

Cragg Wood	1838	Freemans Journal 24/8/1838
Corran Tuathail	1873	Freemans Journal 22/9/1923
Rosscarbery	1893	Southern Star 11/2/1893
Macroom	1902	Southern Star 19/7/1902
Bandon	1925	Irish Independent 26/6/1925
Ogonnelloe	1938	Nenagh Guardian 17/12/1938

"I shot what I took to be a Wild Cat at Liss Ard, Skibbereen, during "the winter of 1873-74, probably in January 1874. The place was high, rocky ground, on the skirt of a young plantation. I just got a glimpse of it passing through the gorse and brambles and thought it might be a Marten Cat. We were beating for Woodcocks. The retriever fetched it, and when she came out of the covert the Cat had her by the nose. [1]

(* Also known as the *Irish Naturalists Journal*.)
The table, by the way, shows reports of Irish wildcats over a period of 100 years.

There is an article in the *Freemans Journal* in 1923 which gives the etymology and history of the Wild Cat in Ireland along with some other Irish animals:

FURTHER LOCAL NAMES DERIVED FROM ANIMALS,MOSTLY WILD

"When the cat appears in place names, it is not the domestic animal that it meant, but the wild cat which was at one time common in Ireland, and was certainly not extinct fifty years ago [i.e. c.1873-R] for I know a man that used to trap them in a wood at the foot of Corran Tuathail,Co Kerry. This wild cat was a fierce brute, called Cat Crainn, tree cat,and occasionally Mada Crainn,tree dog, the latter being also the Irish name of the squirrel.

[So could that offer the exciting possibility of two types of Irish wild cat?- R]

Knockannacuit,parish of Lismore, Co.Waterford is Cnocan-a-Chuit,the Cat`s Hillock. and Meenachuit, Inniskeel,Co.Donegal, Min-a-Chuit,the Cat`s Smooth Spot. In each of these cases the cat appears in the singular number. He comes into place numbers more frequently in the plural. Carnagat, parish of Killevy,Co.Armagh,is Carn-na-gCat, the Cat`s Cairn;Carrignagat, Kilmocomago parish,Co Antrim,has practically the same meaning.Knocknagat, Cuocna-gCat, the Cat`s Hill;Lisnagat,Lios-na-gCat,the Cat`s Liss and Lisheeunagat,Lisin-na-gCat,the Cat`s Little Fort,Lisin being the diminuative of Lios;

No doubt many of these names are legendary. Those who have read Father O`Leary`s "Guaire" will recall how the King of the Cats carried away the haed of the Trom-Daimhe, because , in a fit of temper, he had lampooned the cats for neglecting to kill the mice that had stolen the dainty meal which Guairo had sent him, when the Ard-Ollamh was in a sulky mood and not inclined to eat. [2]

In 1956 a Co.Leitrim newspaper reported a modern Dobhar Chu sighting:

STRANGE MONSTER

The story of a monster in Glenade Lake is again in circulation. According to the account of some young men who claim to have seen the monster while fishing there, it has the head of a hound, the tail of a fish and is about six feet in length. Rumours of this kind excite more interest where Glenade Lake is concerned because of a thrilling scene which took place 236 years ago ,when a Miss McLoughlin known by her maiden name Grace Connolly, was killed by Dobhar-Chu while washing clothes in the lake convenient to her home in Creevelea.[3]

1   *Irish Naturalist* vol. 17 no 7. July 1908 pp 140-141

2 *Freeman's Journal.* September 22nd 1923
3 *Leitrim Observer.* November 10th 1956.

Seeing as we ended on a watery theme, I will end with a watery song;

The Levellers `The Boatman`

If I could choose the life I please
Then I would be a boatman
On the canals and rivers free
No hasty words are spoken
My only law is the river breeze
That takes me to the open seas
If I could choose the life I please
Then I would be a boatman

STILL ON THE TRACK OF UNKNOWN ANIMALS

T he Centre for Fortean Zoology, or CFZ, is a non profit-making organisation founded in 1992 with the aim of being a clearing house for information, and coordinating research into mystery animals around the world.

We also study out of place animals, rare and aberrant animal behaviour, and Zooform Phenomena; little-understood "things" that appear to be animals, but which are in fact nothing of the sort, and not even alive (at least in the way we understand the term).

Not only are we the biggest organisation of our type in the world, but - or so we like to think - we are the best. We are certainly the only truly global cryptozoological research organisation, and we carry out our investigations using a strictly scientific set of guidelines. We are expanding all the time and looking to recruit new members to help us in our research into mysterious animals and strange creatures across the globe.

Why should you join us? Because, if you are genuinely interested in trying to solve the last great mysteries of Mother Nature, there is nobody better than us with whom to do it.

Members get a four-issue subscription to our journal *Animals & Men.* Each issue contains nearly 100 pages packed with news, articles, letters, research papers, field reports, and even a gossip column! The magazine is Royal Octavo in format with a full colour cover. You also have access to one of the world's largest collections of resource material dealing with cryptozoology and allied disciplines, and people from the CFZ membership regularly take part in fieldwork and expeditions around the world.

The CFZ is managed by a three-man board of trustees, with a non-profit making trust registered with HM Government Stamp Office. The board of trustees is supported by a Permanent Directorate of full and part-time staff, and advised by a Consultancy Board of specialists - many of whom are world-renowned experts in their particular field.   We have regional representatives across the UK, the USA, and many other parts of the world, and are affiliated with

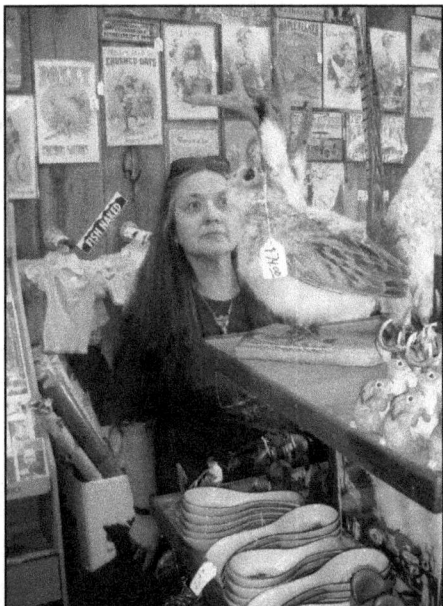

You'll find that the people at the CFZ are friendly and approachable. We have a thriving forum on the website which is the hub of an ever-growing electronic community. You will soon find your feet. Many members of the CFZ Permanent Directorate started off as ordinary members, and now work full-time chasing monsters around the world.

Write to us, e-mail us, or telephone us. The list of future projects on the website is not exhaustive. If you have a good idea for an investigation, please tell us. We may well be able to help.

We are always looking for volunteers to join us. If you see a project that interests you, do not hesitate to get in touch with us. Under certain circumstances we can help provide funding for your trip. If you look on the future projects section of the website, you can see some of the projects that we have pencilled in for the next few years.

In 2003 and 2004 we sent three-man expeditions to Sumatra looking for Orang-Pendek - a semi-legendary bipedal ape. The same three went to Mongolia in 2005. All three members started off merely subscribers to the CFZ magazine. Next time it could be you!

We have no magic sources of income. All our funds come from donations, membership fees, and sales of our publications and merchandise. We are always looking for corporate sponsorship, and other sources of revenue. If you have any ideas for fund-raising please let us know. However, unlike other cryptozoological organisations in the past, we do not live in an intellectual ivory tower. We are not afraid to get our hands dirty, and furthermore we are not one of those organisations where the membership have to raise money so that a privileged few can go on expensive foreign trips. Our research teams, both in the UK and abroad, consist of a mixture of experienced and inexperienced personnel. We are truly a community, and work on the premise that the benefits of CFZ membership are open to all.

Reports of our investigations are published on our website as soon as they are available. Preliminary reports are posted within days of the project finishing.

Each year we publish a 200 page yearbook

We have a thriving YouTube channel, CFZtv, which has well over two hundred self-made documentaries, lecture appearances, and episodes of our monthly webTV show. We have a daily online magazine, which has over a million hits each year.

Each year since 2000 we have held our annual convention - the Weird Weekend. It is three days of lectures, workshops, and excursions. But most importantly it is a chance for members of the CFZ to meet each other, and to talk with the members of the permanent directorate in a relaxed and informal setting and preferably with a pint of beer in one hand. Since 2006 - the Weird Weekend has been bigger and better and held on the third weekend in August in the idyllic rural location of Woolsery in North Devon.

Since relocating to North Devon in 2005 we have become ever more closely involved with other community organisations, and we hope that this trend will continue. We have also

© Underground Images 2007

worked closely with Police Forces across the UK as consultants for animal mutilation cases, and we intend to forge closer links with the coastguard and other community services. We want to work closely with those who regularly travel into the Bristol Channel, so that if the recent trend of exotic animal visitors to our coastal waters continues, we can be out there as soon as possible.

Apart from having been the only Fortean Zoological organisation in the world to have consistently published material on all aspects of the subject for over a decade, we have achieved the following concrete results:

• Disproved the myth relating to the headless so-called sea-serpent carcass of Durgan beach in Cornwall 1975

• Disproved the story

of the 1988 puma skull of Lustleigh Cleave

- Carried out the only in-depth research ever into the mythos of the Cornish Owlman.
- Made the first records of a tropical species of lamprey
- Made the first records of a luminous cave gnat larva in Thailand
- Discovered a possible new species of British mammal - the beech marten
- In 1994-6 carried out the first archival fortean zoological survey of Hong Kong
- In the year 2000, CFZ theories were confirmed when a new species of lizard was added to the British List
- Identified the monster of Martin Mere in Lancashire as a giant wels catfish
- Expanded the known range of Armitage's skink in the Gambia by 80%
- Obtained photographic evidence of the remains of Europe's largest known pike
- Carried out the first ever in-depth study of the ninki-nanka
- Carried out the first attempt to breed Puerto Rican cave snails in captivity
- Were the first European explorers to visit the `lost valley` in Sumatra
- Published the first ever evidence for a new tribe of pygmies in Guyana
- Published the first evidence for a new species of caiman in Guyana

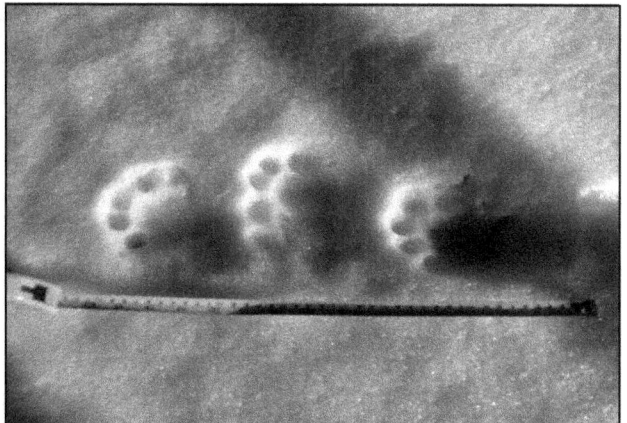

on a monster-haunted lake in Ireland for the first time
- Had a sighting of orang pendek in Sumatra in 2009
- Found leopard hair, subsequently identified by DNA analysis, from rural North Devon in 2010
- Brought back hairs which appear to be from an unknown primate in Sumatra
- Published some of the best evidence ever for the almasty in southern Russia

CFZ Expeditions and Investigations include:

- 1998 Puerto Rico, Florida, Mexico (Chupacabras)
- 1999 Nevada (Bigfoot)
- 2000 Thailand (Naga)
- 2002 Martin Mere (Giant catfish)
- 2002 Cleveland (Wallaby mutilation)
- 2003 Bolam Lake (BHM Reports)

- 2003 Sumatra (Orang Pendek)
- 2003 Texas (Bigfoot; giant snapping turtles)
- 2004 Sumatra (Orang Pendek; cigau, a sabre-toothed cat)
- 2004 Illinois (Black panthers; cicada swarm)
- 2004 Texas (Mystery blue dog)
- Loch Morar (Monster)
- 2004 Puerto Rico (Chupacabras; carnivorous cave snails)
- 2005 Belize (Affiliate expedition for hairy dwarfs)
- 2005 Loch Ness (Monster)
- 2005 Mongolia (Allghoi Khorkhoi aka Mongolian death worm)

- 2006 Gambia (Gambo - Gambian sea monster , Ninki Nanka and Armitage's skink
- 2006 Llangorse Lake (Giant pike, giant eels)
- 2006 Windermere (Giant eels)
- 2007 Coniston Water (Giant eels)
- 2007 Guyana (Giant anaconda, didi, water tiger)
- 2008 Russia (Almasty)
- 2009 Sumatra (Orang pendek)
- 2009 Republic of Ireland (Lake Monster)
- 2010 Texas (Blue Dogs)
- 2010 India (Mande Burung)
- 2011 Sumatra (Orang-Pendek)
- 2012 Sumatra (Orang Pendek)
- 2014 Tasmania (Thylacine)
- 2015 Tasmania (Thylacine)
- 2016 Tasmania (Thylacine)
- 2017 Tasmania (Thylacine)

For details of current membership fees, current expeditions and investigations, and voluntary posts within the CFZ that need your help, please do not hesitate to contact us.

The Centre for Fortean Zoology,
Myrtle Cottage,
Woolfardisworthy,
Bideford, North Devon
EX39 5QR

Telephone 01237 431413
Fax+44 (0)7006-074-925
eMail info@cfz.org.uk

Websites:

www.cfz.org.uk
www.weirdweekend.org

THE WORLD'S WEIRDEST PUBLISHING COMPANY

ANIMALS & MEN

ISSUES 16-20

THE JOURNAL OF THE CENTRE FOR FORTEAN ZOOLOGY

NEW HORIZONS

Edited by Jon Downes

BIG CATS LOOSE IN BRITAIN

PREDATOR DEATHMATCH

NICK MOLLOY

WITH ILLUSTRATIONS BY ANTHONY WALLIS

PHENOMENA

Edited by Jonathan Downes and Richard Freeman

FOREWORD BY Dr. KARL SHUKER

A DAINTREE DIARY

Tales from Travels Daintree tropical North

CARL PORTMAN

THE COLLECTED POEMS
Dr Karl P. N. Shuker

STRANGELYSTRANGE

ly normal

an anthology of writings by
ANDY ROBERTS

HOW TO START A PUBLISHING EMPIRE

Unlike most mainstream publishers, we have a non-commercial remit, and our mission statement claims that "we publish books because they deserve to be published, not because we think that we can make money out of them". Our motto is the Latin Tag *Pro bona causa facimus* (we do it for good reason), a slogan taken from a children's book *The Case of the Silver Egg* by the late Desmond Skirrow.

WIKIPEDIA: "The first book published was in 1988. *Take this Brother may it Serve you Well* was a guide to Beatles bootlegs by Jonathan Downes. It sold quite well, but was hampered by very poor production values, being photocopied, and held together by a plastic clip binder.

In 1988 A5 clip binders were hard to get hold of, so the publishers took A4 binders and cut them in half with a hacksaw. It now reaches surprisingly high prices second hand.

The production quality improved slightly over the years, and after 1999 all the books produced were ringbound with laminated colour covers. In 2004, however, they signed an agreement with Lightning Source, and all books are now produced perfect bound, with full colour covers."

Until 2010 all our books, the majority of which are/were on the subject of mystery animals and allied disciplines, were published by `CFZ Press`, the publishing arm of the Centre for Fortean Zoology (CFZ), and we urged our readers and followers to draw a discreet veil over the books that we published that were completely off topic to the CFZ.

However, in 2010 we decided that enough was enough and launched a second imprint, `Fortean Words` which aims to cover a wide range of non animal-related esoteric subjects. Other imprints will be launched as and when we feel like it, however the basic ethos of the company remains the same: Our job is to publish books and magazines that we feel are worth publishing, whether or not they are going to sell. Money is, after all - as my dear old Mama once told me - a rather vulgar subject, and she would be rolling in her grave if she thought that her eldest son was somehow in `trade`.

Luckily, so far our tastes have turned out not to be that rarified after all, and we have sold far more books than anyone ever thought that we would, so there is a moral in there somewhere…

Jon Downes,
Woolsery, North Devon
July 2010

CFZ PRESS

CFZ Press is our flagship imprint, featuring a wide range of intelligently written and lavishly illustrated books on cryptozoology and the quirkier aspects of Natural History.

CFZ Classics is a new venture for us. There are many seminal works that are either unavailable today, or not available with the production values which we would like to see. So, following the old adage that if you want to get something done do it yourself, this is exactly what we have done.

Desiderius Erasmus Roterodamus (b. October 18th 1466, d. July 2nd 1536) said: "When I have a little money, I buy books; and if I have any left, I buy food and clothes," and we are much the same. Only, we are in the lucky position of being able to share our books with the wider world. CFZ Classics is a conduit through which we cannot just re-issue titles which we feel still have much to offer the cryptozoological and Fortean research communities of the 21st Century, but we are adding footnotes, supplementary essays, and other material where we deem it appropriate.

http://www.cfzpublishing.co.uk/

Fortean Words is a new venture for us. The F in CFZ stands for "Fortean", after the pioneering researcher into anomalous phenomena, Charles Fort. Our Fortean Words imprint covers a whole spectrum of arcane subjects from UFOs and the paranormal to folklore and urban legends. Our authors include such Fortean luminaries as Nick Redfern, Andy Roberts, and Paul Screeton. . New authors tackling new subjects will always be encouraged, and we hope that our books will continue to be as ground-breaking and popular as ever.

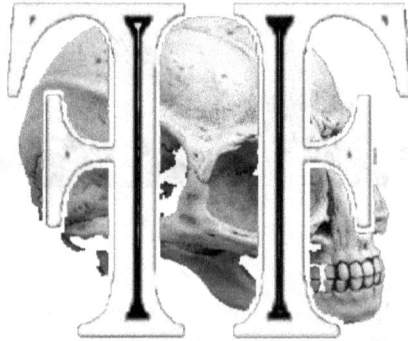

Just before Christmas 2011, we launched our third imprint, this time dedicated to - let's see if you guessed it from the title - fictional books with a Fortean or cryptozoological theme. We have published a few fictional books in the past, but now think that because of our rising reputation as publishers of quality Forteana, that a dedicated fiction imprint was the order of the day.

http://www.cfzpublishing.co.uk/